Why Readers Are Raving About
Becoming Your Dream

"*Becoming Your Dream* is a lifejacket for anyone who feels like they are drowning in their limiting beliefs about their lives. Joan McManus will soon have you swimming laps around the narrow world you previously lived in. Don't miss this opportunity to revolutionize your life."

> — Susan Friedmann, CSP, Author of *Riches in Niches:*
> *How to Make It BIG in a Small Market*

"In *Becoming Your Dream*, Joan McManus debunks the myths that keep people stuck in limiting beliefs. We all have truths to live, and with the tools McManus offers, we can find the courage to learn how to live those truths through the many large and small daily choices we make."

> — Nicole Gabriel, Author of *Finding Your Inner Truth*
> and *Stepping Into Your Becoming*

"Joan McManus has a knack for making people believe in themselves. In *Becoming Your Dream*, she offers ten proven steps to formulate, test, image, believe, take action, and receive your dream. It's a surefire way to create your own destiny."

> — Patrick Snow, Publishing Coach and International Best-Selling
> Author of *Creating Your Own Destiny* and *Boy Entrepreneur*

"If ever there was an empowering story about the power of visualization, it is Joan McManus's story of how she overcame polio, and that's just one of numerous stories and helpful tools she offers in *Becoming Your Dream*. After reading this book, you'll discover how you can use your mental powers in ways you never imagined to bring about your ideal life. Joan is living proof it works."

> — Tyler R. Tichelaar, PhD and Award-Winning Author of
> *When Teddy Came to Town*

"Joan McManus's proven system of transformation in *Becoming Your Dream* has changed my life and moved me forward in ways I never imagined were possible. She taught me how to access and activate powerful mental faculties I already possess, although I was unaware of their ability to help bring my dreams to fruition. I enthusiastically recommend her book, and also recommend Joan McManus as a life-changing coach and mentor."

— Summer DK Smith, CNHC, CPT, PES

"Why should you read *Becoming Your Dream*? Frankly, it's to your cognitive, emotional, spiritual, and financial advantage! It's a can-do template to "become your dream." It's a *bam-bam, it's not that hard, I'll tell you how* guide based on research and testaments from noted persons from yore to now. Joan relates much of her own story, fraught with illness, hardships, and disappointments to illustrate the truths she had to learn about herself to touch and become her dream. She has shared her learnings in a compelling, easily readable format for how to actualize our individual wonders with the power of our minds, values, and intentions. 'By universal law you will become one with your vision, seeing results even beyond your wildest expectations,' she assures her readers. I read Mcmanus's book. I believe her assurances and guidelines. I look forward to any other book she might write from her A to Z studies, experiences, and achievements."

— JoAnn McKinzey, Political Activist, Former Educator and Teacher Trainer, K-University, College of Education

"In *Becoming Your Dream*, Joan McManus brings together ancient wisdom, principles of leadership, and scientific studies from various fields to illustrate the principle that each of us has the right and responsibility to live a joyful, empowering, fulfilling life. The concrete steps are here to help anyone transform their life into one they truly love."

— Candy Glade, Educator, Teacher Trainer

"Joan McManus and I have been close friends for more than fifty-five years, and I have witnessed and/or participated in many important phases of her life. I can, therefore, attest that she truly has believed in and achieved her dream. Joan's book has specific exercises and examples for making one's dream a reality. It is well written and worthwhile."

— Betsy H. Birkett, Friend, Sorority Sister,
Kappa Kappa Gamma

"*Becoming Your Dream* is a must read for anyone who is ready to make quantum leaps forward into a life they would love living! Whether you are feeling stuck or are having lackluster results with your current health and wellbeing, relationships, vocation, or time and money freedom—this book has you covered. Joan masterfully teaches you proven principles that will help you transform your life and accelerate your results. Let this book inspire you to live your best life!"

— Anne L. Prinz, Author of *Living
Your Exclamation Point Life!*

"Joan reveals the framework of a proven process for building dreams, and provides illuminating quotes and stories from ancient and modern sages who inspire and motivate, as well as her own personal stories of life lessons. She offers strategies for action that when implemented will propel dreams and goals into manifestation. You will benefit from the wisdom and strategies presented in this book."

— Nancy Holley Hood, Business Owner, Certified Dream
Builder Coach, Author, Seminar Leader, and Inspirational
and Educational Adventure Retreat Leader

"A veritable how-to guide, Joan McManus's *Becoming Your Dream* offers a powerful blend of wisdom of the ages along with her own extensive experience shepherding others along this sacred path to full expression of themselves. This wonderful read will leave you feeling exhilarated, joyful, and more resourced to navigate your own calling. Thank you, Joan, for sharing your embodied wisdom with the world!

— James Woeber, CEO and Co-Founder
of Art of Heartful Living Leadership and Human
Potential Training and Consulting,
www.ArtofHeartfulLiving.com

A PROVEN SYSTEM TO CREATE ABUNDANT HAPPINESS, HEALTH, AND WEALTH

BECOMING YOUR
Dream

HARNESS YOUR SIX HIDDEN SUPERPOWERS
TO TRANSFORM YOUR LIFE

JOAN C. MCMANUS

AVIVA
PUBLISHING
New York

Published by:
Aviva Publishing
Lake Placid, NY
(518) 523-1320
www.AvivaPubs.com

Joan C. McManus
Joan@JoanMcManus.com
JoanMcManus.com

ISBN: 978-1-947937-60-4
Library of Congress Control Number: 2018911327

Editor: Tyler Tichelaar, Superior Book Productions
Cover Design and Interior Layout: Nicole Gabriel, AngelDog Productions

Every attempt has been made to source properly all quotes.
Printed in the United States of America.
First Edition

TO DORA MARGARITE MOELLER
AKA GRANDMA CLARK

Dora was an urban pioneer woman who became manager of a flagship furniture store in Los Angeles—an unheard of accomplishment in 1910. The daughter of German immigrants, Dora moved from the small farm in Iowa to pursue her dream. She later became successful in commercial real estate and moved her parents and sister to her beloved California.

Grandma Clark was the only mother I knew until I started school. She taught me the principles we'll study in this book—although it took me years and years of study with many masters before I even came close to her level of understanding and faith. Giving me the foundation for a happy life, Grandma Clark continued to support me mentally and spiritually long after I had to leave her nurturing home.

It is now my mission to share the wisdom she worked so hard to impart to me and to all who were fortunate enough to have her in their lives. In that spirit, I have done my best to put in writing the Truths that will allow you to use the talents your Creator gave you so you could have everything you long for—to become the person living your finest dream.

CONTENT

EXPERIENCE JOAN MCMANUS'S TRANSFORMATIVE COACHING

As a certified life and business success coach, Joan McManus delivers a proven coaching model that has transformed the lives of thousands of people. While bringing her passion for demystifying the universal laws that each of us lives by (whether we recognize it or not), Joan also includes her science and research background to help each individual better apply the most effective performance improvements.

Tailoring her coaching sessions to individual needs, Joan supports each client in creating a clear vision of his or her goal, and guides in creating action steps to implement at once. Receiving motivation, support, accountability, and increased awareness from Joan, clients progress quickly in navigating the gap between where they are now and where they want to be.

As an entrepreneur and former organizational leader, Joan brings her own experience to the client. She gives her coachees the tools to grow, expand their own consciousness, and stand on the edge of their comfort zone to be able to stretch beyond their current boundaries.

Because Joan firmly believes in the greatness potential of each of her clients, she forms a strong bond with her clients. They report becoming passionate and confident about achieving the lives they would love to live, and then they take the necessary steps to create those lives.

Contact Joan. She will help you discover how the life of your dreams is waiting to emerge within you.

To set up a complimentary, no obligation, thirty-minute consultation by phone, email Joan at Joan@JoanMcManus.com or text her with your name and time zone at 214-793-3203.

"Dream as if you'll live forever. Live as if you'll die today."

— James Dean

INTRODUCTION:

EVALUATING YOUR LIFE NOW

"Whatever you can do, or dream you can, begin it.
Boldness has genius, power, and magic in it."

— Johann Wolfgang von Goethe

Do you have a personal longing for something or someone? A desire you have had for a long time, but it seems impossible to possess? Is it tucked in the recesses of your mind so that you won't seem foolish for wanting it so?

Do you live an exciting, interesting life? Do you get up in the morning looking forward to the day's adventures? Do your daily experiences keep you at your growing edge, or have you become a slave to daily routines that dull your senses and sap your energy? Are you surrounded with people who encourage you to try new things? People who keep you laughing? Who inspire you?

What about people, situations, or things you would prefer less of in your life? Are there personal contacts who bring you down? Do you have some things in your home that, instead of bringing you joy, give you "the blahs"? Is there something you purchased thinking, "This'll do," yet it let you down due to poor quality?

How do you satisfy your need to grow? Do you give yourself permission to push the envelope? Do you dare to express yourself creatively, or do you listen to others' opinions before stepping forth?

Are you living a life of abundance? Do you have enough money to meet your needs, satisfy your wants, and give to loved ones and your community in ways that make you happy?

Or are you living from the worldview of scarcity? Worried that there won't be enough to pay the bills, to purchase things to surprise and delight friends, or to donate to the causes you believe in?

What if you had a time-tested, proven blueprint to change your life into one you would truly love? A full spectrum life that includes great health, true wellbeing, fulfilling relationships, a purpose-driven, passion-filled vocation, and time and money freedom?

That dream life *can* be yours!

The process to achieve it is spelled out for you in this book. What is required for you to reap the benefits is that you come with an open mind, ready to play full out.

Bring your confidence because this blueprint has worked for thousands of people, and it *will* work for you. Best of all, it is relatively easy to follow when you take small steps each day toward your vision.

How do I know this process will work? I've used it myself, I've seen it work for those I've coached, and I've seen the success of hundreds I've personally met who have used this technology.

One of my biggest goals became a reality when many of my colleagues were retiring. After five years of academic and leadership training, I was named school principal. My varied professional experience in education, speech and language development, psychology, and finance served to give me a broad perspective as a school administrator. As I inserted my key to unlock my office each morning, I often thought, "I am so happy! Thank you for this day in my school." With a creative and talented staff, professional growth was fun. Teachers, students, staff, parents, and community became like family. Therefore, when it came time for me to retire, I was left with a huge chasm. My personal identity was enmeshed in my professional role. After a year of rejuvenating and playing, I was faced with a life crisis: I found myself asking these three questions: Who am I now? What do I want? How can I serve? I had long before discovered that serving was my "high."

Imagine my surprise when, during a meditation training, my instructor, Dr. Deepak Chopra, offered these same questions! He also added a fourth valuable question: What am I thankful for? I was left wondering, *Are these universal questions each of us must answer in our individual way, depending on our own life's path?*

I discovered the common thread in my diverse professional career that had made me euphoric was coaching others' growth. Having been a personal growth student myself for four decades, I was alert to the expansion that manifested in many ways in others' lives. Hence, I invested the time and resources in a premier training institute to become a Certified DreamBuilder Coach and Life Mastery Consultant.

As I coached clients, I discovered many others who struggled with establishing a new identity after a major life change. Divorce, retirement, "empty nest syndrome," corporate restructuring, or even reaching a plateau in career development can create a tremendous void. The "What now?" feeling, the "Is this all there is?" concern, or the "When is it time for my voice to be heard?" or "Do I want to be heard?" conundrum was common for all of us. Asking these questions can bring formerly successful women and men to their metaphorical knees. Conditioned beliefs can hinder our freedom to explore the question, "What do I really want now?"

I continue to be surprised that most clients, when they begin to design their ideal lives, after some coaching and deep searching within, end up fine-tuning or even eliminating several of their initial choices and directions. That transition is positive, and it illustrates that, influenced by the way others expect us to live or work, we often don't have an awareness of our own unique gifts or our deepest desires. As we expand our consciousness, we discover the talents we can bring to the world in our own spicy way. Many of us have not asked our hearts, "What do I really, really want?"

In this book, we will explore the tools needed to answer that empowering question. We will travel the proven steps many serious dreamers have taken to live a life of happiness, purpose, and abundance. The world needs each of us now, more than ever, in our special uniqueness.

A window of opportunity is open now, one greater than ever before in the history of mankind. As Eben Pagan, renowned business coach, keynote speaker, and best-selling author, illustrates in his book *Opportunity*, the spread of knowledge and technological innovation has been dramatically accelerated by the internet. "Change is now reaching into every domain, and it will transform lives dramatically. Your domain, whatever it is, will be transformed. This is creating the biggest explosion in opportunity that humans have ever seen."[1]

As the flip side of this picture, Pagan shares Margaret Atwood's warning that potential has a finite shelf life. In the future, the windows of opportunity will open and close faster with the "intersection of idea and multiple enabling innovations."[2]

Price Pritchett, PhD and business consultant, states this trend another way in his book *You²*. He teaches that we don't have to settle for our current circumstances, and says further: "Right now, in this moment, you are capable of *exponential improvement* in your performance. You can *multiply* your personal effectiveness, hit new highs, and shatter your old achievement records. The results you can have will be hard for you to imagine."[3]

Pritchett promises a completely different plane of success, a breakthrough experience. Not only can our level of performance im-

---

1 Pagan, Eben. *Opportunity*. p. 28-9.
2 Ibid. p. 29.
3 Pritchett, Price. *You²*. p. 2.

prove, but our rate of accomplishment can soar. All of this can happen with far less effort than our past attempts. We don't have to be content with improving our lives incrementally, gradually.

We must reframe our ideas about the universe and how we fit into it. This shift requires a major rethinking of such concepts as time and space and how human consciousness operates. What was once considered "woo-woo" and "otherworldly" has now been found to be accurate by physics, psychology, biology, neurology, and philosophy. Ancient wisdom's truths are now resurfacing on the planet; they are being better understood by those willing to open their minds and hearts to new possibilities.

If you have not already seen the work of Fred Alan Wolf in his YouTube videos, I encourage you to watch the Dr. Quantum series, particularly the Famous Double Slit experiment. As Dr. Quantum, Wolf gives a clear explanation of life at the subatomic level.

This background information will support your journey through the ten steps to become your dream because it will help you understand that we live in an energetic field that responds to the vibrational influences we emit from our mind and body. As you watch these enlightening videos, be sure to note that particles make quantum "jumps" without apparent effort and without covering all the bases between start and end points. But these quantum jumps require a radical departure from some old habits, a departure that can feel awkward and risky at first. However, if you are willing to employ the new mindsets and behaviors we'll cover in this book, you will become you[2]!

Just as physical boundaries, once broken, become accessible (think the four-minute mile), the limiting paradigms (belief models) we were given by well-intended, but misinformed parents, teachers, peers, and society-at-large can be replaced with empowering beliefs and habits. To help you know in your core that a transformation is not only possible, but probable—and to be expected—for you, I'll relate stories throughout of people just like you and me who used these same principles for success.

Are you ready to defy common sense and move into a higher orbit, skipping several rungs? Quantum leaps, once taken, prove to be simple, energy-efficient, and time-saving. *The time has come* to employ our Higher Order Thinking Skills (HOTS) and use proven steps to make a quantum leap!

Join me to conceive and achieve the life of your wildest dreams! The journey is easier when you stay focused and surround yourself with like-minded people. It is also great fun when you remain light-hearted as you discover what you are meant to do now.

By studying the proven ten steps to becoming your dream, you will learn:

1. The Higher Order Thinking Skills (HOTS). Think of them as the six mental superpowers you must harness.
2. How to design your dream.
3. How to test, digest, and invest in your dream.
4. How knowing *who* and *where* you are solidifies your belief and fortifies your faith.
5. The ultimate power of your thoughts and words.
6. How to get in motion to make things happen.
7. How to face challenges.
8. Ways to examine and upgrade your current models and systems and to garner support.
9. The most effective ways to weed and fertilize the garden of your mind to nurture your dream.
10. How to live in your abundance.

You decide….

When is the time?

Now.

Who can do it?

You can.

Find your deepest *why.*

Ask yourself, "*Why* must I do this?"

Once you have the answer, holding your *why* in your heart,

Do it!

Do what you need to do to become your dream!

I'll be here to help you along the way. Enjoy the journey. I believe in you. Now turn the page and let's begin.

Joan C McManus

Exercise

Reflect briefly on the following points. Write your personal responses spontaneously and effortlessly. In other words, write without overthinking. This beginning is for your eyes only.

1. If time and money were no issue, what is one vibrant role you would love to assume at this time in your life?

_____

_____

_____

_____

_____

_____

2. Describe a talent of yours that you want to develop and express more fully.

_____

_____

_____

_____

_____

_____

3. Name one habit you would like to change because it no longer serves you.

_____

_____

_____

_____

_____

_____

"Do not go where the path may lead, go instead where there is no path and leave a trail."

—Ralph Waldo Emerson

PART I

CREATING YOUR DREAM

To ensure you create the greatest dream possible, it's best that you have a clear understanding of the magical mental faculties gifted to you at birth. These talents have been active in your life, possibly without your direct awareness. They may be helping you achieve goals, or they could be sabotaging every effort you make to change course in life. Whatever the case, they are far too valuable to continue playing an important role in your results unless the results are exactly what you would design. So, in Chapter 1, we'll examine the Higher Order Thinking Skills we each have been given as a precious commodity for co-creating an ideal life. You'll learn what they are and how you can tune in to use them to full benefit.

In Chapter 2, we'll progress to the actual process of designing the life you want. Not just a life you think might be possible given your current circumstance, but the life you dream of having.

In Chapter 3, we'll test that dream, prodding you to see if it's big enough to warrant your loving work to make it a material reality. You'll put it on to see if it actually fits, and you'll make the commitment to have it.

The circular staircase pictured below is a metaphor for the process we'll begin. It depicts the access to my favorite rooftop view of the magnificent sunset on the Pacific Ocean. As you can see, the steps are on a steep incline. However, the ascent is made much easier by the sturdy handrails that guide you up the stairs. Each of the steps represents a proven step necessary to reach a lofty goal. The sturdy handrail and center pole, when grasped firmly, make the climb so much easier, steadying the climber at each new level. I compare these supporting structures to the Higher Order Thinking Skills that make the rise to the new height not only possible, but also enjoyable.

The comparison of the journey we are about to take to this circular staircase doesn't end with the steps, rail, and pole. When you do reach the top, a whole new vista will be available to you.

Just as we'll find on our climb together, when one plateau is reached, the new view expands our horizons, and our dream grows. That is the beauty of this ever-upward cycle; it becomes more exciting with each small step we take, and it leads to a life so fulfilling that it is even hard to imagine it now.

Let's dive in to learn about the Higher Order Thinking Skills (HOTS) we want to use fully on this journey, and then go forward to build and test the dream we are going to become!

"You have power over your mind—not outside events. Realize this, and you will find strength."

—Marcus Aurelius

CHAPTER 1

IGNITING YOUR HOTS

"What lies behind us and what lies before us are tiny matters compared to what lies within us."

— Ralph Waldo Emerson

To make meaning of the world around us, you and I have been given five senses. Through seeing, hearing, smelling, touching, and tasting, we are constantly bombarded with sensory data that gets interpreted by the conscious and subconscious mind to help us survive and thrive.

We also have six magnificent superpowers we employ, but possibly not with the conscious precision needed to gain their full benefit. These six invisible mental faculties shape our thoughts, emotions, ideas, beliefs, actions, and finally—our results. I'd like you to review that statement again: These six mental faculties ultimately determine the life we live. For this reason, I call them our Higher Order Thinking Skills, using the simple acronym HOTS.

You and I use our HOTS every day, but often with little focus toward achieving our desires.

The Higher Order Thinking System consists of these six skills:

1. Imagination
2. Intuition
3. Will
4. Memory
5. Reason
6. Perception

Let's now look at a brief functional definition of each one. Then, for better understanding, we'll apply them in a true story.

Imagination

> "Imagination is more important than knowledge. For knowledge is limited, whereas imagination embraces the entire world, stimulating progress, giving birth to evolution."
>
> — Albert Einstein

Eben Pagan, in his book *Opportunity*, points out that we now live in a "designer world," one created by the imagination of those who wanted to optimize our comfort, safety, and convenience. For example, our entire physical habitat has been shaped in some way by human hands, changing our existence from living in the wild to "living within multiple layers of purpose-built environments." He says that is also true of our experiences, relationships, thoughts, and social groups—that the imagination of man has influenced every aspect of our lives, and that this shift is happening at faster and faster rates. "The gap between imagining something in our mind and then creating it in the real world is shrinking every day."[4]

We can look at the advances in technology to see the almost unfathomable rate of change. Ray Kurzweil, the brilliant futurist, says we are moving rapidly toward an "imagine it and create it" reality—a place where we envision what we want in our minds, and then we use knowledge, technology, and other resources to manifest it by shaping, designing, and creating the environment around us. By studying

---

4 Pagan, Eben. *Opportunity.* p. 8.

the increasing speed and decreasing cost of computers over the last hundred years, Ray Kurzweil found a pattern of double exponential growth and change. Looking deeper, he found that intelligent life and processing power have been accelerating since life began.

Our minds cannot easily grasp exponential change, so Eben Pagan gives the lotus pond example, one I have often used to teach exponents to math students. Imagine a pond where lotus plants grow. On Day One, there is a single lotus plant. If the number of plants doubles each day, and it takes thirty days for the pond to be completely covered, what will you see as the process continues? "On Day 25, only about 3 percent of the pond is covered. On Day 27, only 12 percent of the pond is covered. But then on Day 29, it's half the pond. Then on Day 30, it's the entire pond."[5] For most of the month, you would not have noticed anything happening. Then, suddenly—"out of the blue"—the pond is covered in lotus blossoms. (Think back to this metaphor as you employ the ten steps to becoming your dream.)

Eben Pagan takes the futurist's study of growth to another level by saying that Kurzweil is referring to something fundamental about the nature of the reality we live in right now—right at this moment—and evident since the beginning of life:

"We were born into a world that seems to want us to create what we first imagine in our mind. It is a reality that is spectacularly friendly to creative design and manifestation."[6]

Pagan calls this reality the Creativerse (creative + universe), going further to say that in the Creativerse, "we participate in the creative design of reality itself."[7]

Take a breath here and reread that last point. What power and dominion that idea conveys to the individual who engages in purposeful imagination! But to reap the benefit of this truth, you and I must believe that the opportunity for transformation can be created. As

---

5 Ibid. p. 11
6 Ibid. p. 9.
7 Ibid. p. 10.

Pagan says, "You must be convinced that you have the power to shape and create the various aspects of your life."[8] You also must understand that the more creative you are in shaping your environment, the more imaginative and creative you will become because it is a virtuous cycle.

Pagan points out that this is the way the world works, and if you don't step up to create your own dream, you will find yourself living in a world created by others, one that is not necessarily a good fit for your ideal life. "Those of us who don't know how to imagine and create our own environments, relationships, thoughts, feelings, and communities will be less able to cope and thrive, while those that have developed this mindset and skill will be at a huge advantage."[9]

My Wage

I bargained with Life for a penny,

And Life would pay no more,

However I begged at evening

When I counted my scanty store.

For Life is a just employer,

He gives you what you ask,

But once you have set the wages,

Why, you must bear the task.

I worked for a menial's hire.

Only to learn, dismayed,

That any wage I had asked of Life,

Life would have willingly paid.

This familiar poem by J. B. Rittenhouse speaks to the point—you and I must boldly fire up our imaginations to bring into focus the biggest, most outrageously wonderful dreams we will work to be-

---

8 Ibid. p. 10.
9 Ibid. p. 10.

come. When I asked my brother, a highly successful businessman, what he thought of a particular figure I was envisioning to earn in an upcoming venture, he looked at me puzzled. "Why wouldn't you go for ten times that amount? The work involved to achieve the goal is the same."

Let's not be hindered by our ideas of what is possible based on our current or past circumstances. Instead, we can seek new heights, using our imagination as the first step in envisioning our transformed life.

There are two forms of imagination: synthetic (the most common) and creative (used by geniuses and those daring enough to step into the unknown). Synthetic imagination is used to upgrade an existing concept, such as converting a bicycle into a motorcycle. Creative imagination produces structures, objects, or concepts heretofore believed impossible. The airplane and phonograph are prime examples of creative imagination. Napoleon Hill gives a clear description and expands the applications of these two types of imagination in his book *Keys to Success*.[10]

The biggest obstacles to the creative process for these life-changing creative inventions has been—up until now—the physical time, money, or people it would take to effect the change. But in the near future, not time, not money, and not even other people will be the limiting factors! "Then the biggest bottleneck will be our imaginations."[11] It is critical that we begin now, while we have the opportunity to move at a human pace, to develop our imaginations and our designing skills! We still have the chance to practice in slow motion those things that will be happening soon at increasing speeds. By imagining a change in our lives, and then making it happen, we can prepare ourselves for the abundant reality that is becoming more and more visible each day as we move into this time Pagan calls "The Great Acceleration."

---

10 Hill, Napoleon. *Keys to Success*. p. 165-172.
11 Pagan, Eben. *Opportunity*. p. 10.

Often, I hear people say they have no imagination, but I never believe them. Each time one of us worries or feels anxious about a possible outcome, we are using our imagination—imagining a possible future result—but one that may not occur. Mark Twain is reputed once to have said, "I am an old man and have known a great many troubles, but most of them never happened." With practice, each of us can strengthen our imagination to neutralize our fears and act productively.

Imagination is a skill you can consciously develop. One way is to observe two objects and contemplate how they might be connected. For example, decide what they have in common. Or envision a third object that could connect them. What do your iPhone and a candle have in common? Perhaps they both provide light or put out heat. A car and a tree might be connected by both possibly consisting of wood, or both utilizing round shapes. Using your mind to connect and combine things in new ways and transfer what you learn in one domain to another seemingly unrelated situation is becoming more and more important.

The physicist Richard Feynman applied his observation of a spinning plate in the air to discover how electrons spin around the nucleus in an atom. Bill Bowerman, a track and field coach, applied the pattern of the waffles he had for breakfast to create the Nike Waffle Trainer with soles that worked on different types of surfaces. Steve Jobs transferred the spatial design he learned in a calligraphy class to the Macintosh.

We will become even more powerful in our use of imagination by learning to create a crystal clear image in our minds, a thought that then becomes an energetic "thing" and initiates a chain of emotional, chemical, and physical reactions that ultimately change our current reality. (More later on this important aspect.)

Intuition

The HOTS' intuition can be highly developed to connect you and me to the world of Infinite Intelligence, a domain Thomas Edison called the Land of Solutions. Each of us has experienced that "still, small voice," but many of us don't heed it, either because we have not learned to distinguish it from the din and clatter of daily life and our racing, random thoughts, or because we don't trust its message.

Years ago, after purchasing an antique dining suite, I looked at the china closet and told my husband, "I really want that to be filled with fine china," a seemingly foolish remark, since I had no prospect of shopping for tableware. A couple of hours later, our doorbell rang, and there stood my husband's elderly Uncle Pat, whom I had never met. Uncle Pat was carrying a large box, and said a little sheepishly, "I've had this china in my garage for over twenty years, and I just had the strongest inclination that I should give it to you. It occurred to me that it might mean something special to Clark, since the family used it every Sunday." Carefully opening the cartons, we found hand-painted Bavarian china, exquisitely crafted by my husband's grandmother. It was truly the most elegant china I had ever seen, with unique pieces no longer found in modern sets. With no provocation from us, Uncle Pat listened to his intuition and brought us a gift that provided years of enjoyment as extended families continued to feast on the true pieces of art.

So often we are given this opportunity to expand the love we have enjoyed ourselves, and many times this requires listening to the still, small voice of intuition. With practice using techniques we'll cover in later chapters, you will become fluent in your communication with your own intuition, and that is when you will become unstoppable!

Will

The HOTS' *will* is the ability to focus—the discipline to hold your attention on your desired results, despite bombarding distractions. This mental function is very different from the struggle of using willpower. Willpower involves force and requires excessive energy. It is not sustainable, as clearly pointed out in Benjamin Hardy's book, *Willpower Doesn't Work*. Hardy explains that, according to psychological research, willpower is like a muscle, a finite resource that depletes with use. At the end of a taxing day, the individual is left "naked and defenseless," with no control to stop the targeted behavior.[12] Hardy teaches that willpower is only required because of the following deficits:

- Not knowing what you want
- Your desire (your *why*) isn't strong enough
- You aren't invested in yourself or your dreams
- Your environment does not support your goal

In stark contrast to willpower, will is a calm, clear, prioritized positioning of one's attention on taking action steps toward a vision.

After an unsuccessful climb, Sir Edmund Hillary, the mountaineer, promised his adversary, Mount Everest, "I will come again and conquer you because as a mountain you can't grow, but as a human, I can." In 1953, his will prevailed, and he became the first man to reach the peak.

Hardy reveals what is necessary for success: making a committed decision and ensuring focus—in other words, using will—instead of willpower (which can only yield minimal progress). He gives the following actions, which are great examples of proactively, consciously using will:

- Investing up-front to surround yourself with successful, inspiring, and motivational people
- Making your goal public to supporters
- Setting a timeline for action and completion
- Installing several forms of feedback and accountability, both long- and short-term
- Removing everything in your environment that opposes your goal and creating a supportive external space

Thomas Edison said, "Many of life's failures are people who did not realize how close they were to success when they gave up."[13] The best example I know of someone who kept her will to succeed is my young-

---

12 Hardy, Benjamin. *Willpower Doesn't Work*. p. xii.
13 *The Edison and Ford Quote Book*. p. 17.

er daughter during her quest to own a horse. At age ten, Paige already demonstrated Edison's belief that "the most certain way to success is always to try just one more time." Using the money she'd earned doing chores in the neighborhood and the allowances she'd saved, Paige began taking horseback riding lessons. By the time she was twelve, working in her spare time cleaning houses or doing yard work, she had saved $400.00 (in addition to the money for riding lessons). She became a regular at the riding stable, willing to do any work at hand. Horse owners eventually trusted her to walk and lunge their valuable horses when they traveled.

Eager to see her funds grow, Paige asked me to put her money to work in the stock market. I declined, telling her I was an investor, not a "day trader." She implored me to help her. Although I was outside my area of expertise, thinking I would keep a close watch on that small purchase to detect unfavorable movement, I reluctantly acquiesced. In a few months, there was a takeover of "her" company and an unheard of appreciation of the stock. Soon, Paige had $2,000.00! To quote Thoreau, this was "a success unexpected in common hours." The horse she had found through her stable connections was for sale. The price of this well-trained horse was $5,000. Paige's negotiations—at age thirteen—were so effective that the owner decided to reduce his asking price by 60 percent. Paige became the proud owner of a beautiful quarter horse named Copper.

This synopsis doesn't do justice to the many obstacles this young girl faced. One of the primary challenges was convincing her parents, who had clung to the false assumption that she was just going through a phase and would soon lose interest. In retrospect, of course, I wonder, "What were we thinking?" However, the journey for this young hero required taking one small step at a time, and then self-correcting when she found herself off track, until she eventually became a partner with her beloved Copper. Persevering through the process paid huge dividends for her in the development of her HOTS, which she put to use later in discovering her chosen path in life.

During her three-year doctoral program in chiropractic studies, Paige developed a life-threatening bone infection following major knee surgery. Despite medical advice to drop out of school, her "can do" faith kicked in. I still shudder when I remember the image of her working on a cadaver in anatomy lab from her wheelchair—with one leg extended in a splint and a powerful antibiotic running into her veins from an IV! Not only did Paige complete her training, but she managed a sweep of the majority of the honors and awards given at graduation for academics and service. She now has the tenacity, imagination, courage, and intelligent confidence to go for the gusto in other aspects of her life. Because she learned to harness her Higher Order Thinking Skills over a sustained period of time to achieve a goal she was passionate about, she became an unstoppable force in achieving her dreams.

By using one's will to focus on the goal, there are, as Henry Ford said, no dead ends. There is always another way. "The first requisite for success is the ability to apply your physical and mental energies to one problem incessantly without growing weary." German author Johann Wolfgang von Goethe taught that "Things which matter most must never be at the mercy of things which matter least." We heed Goethe's advice by keeping our focus on our vision, and redirecting it when we stray, thus using our *will* positively.

In his bestseller *The One Thing*, Gary Keller, chairman and co-founder of Keller Williams Realty, Inc., explains that will is also the skill of prioritizing, focusing on those things that have meaning and purpose for you, and being certain of those things you will not do. He says the first step to extraordinary results is to understand the concept of the *one* thing and believe it can make a difference in your life. Use will to maintain your focus and consciously choose what you will allow from your outer world to enter your inner world, for our inner and outer worlds play off each other.

Wallace Wattles sums it up nicely for us in his book *The Science of Getting Rich*. "Use your mind to form a mental image of what you want, and to hold that vision with faith and purpose; and use your will to keep your mind working in the Right Way."[14]

---

14 Wattles, Wallace. *The Science of Getting Rich*. p. 33.

Memory

Contrary to our social conditioning, everyone has a nearly perfect memory. Our retrieval mechanism may be underdeveloped so that it seems we have forgotten something, but actually that person, experience, or object still exists in our memory. Many memory techniques exist to help you become empowered to remember anything important to you. Harry Lorayne is recommended by one of my mentors, Bob Proctor, as an excellent memory coach. I have used Jim Kwik's programs to improve memory and its offspring, learning. For our purposes of becoming our dream, it is necessary to recognize that memory can work for both the past and the future! We will practice, later in this book, using our memory in both directions. Remember the White Queen in Lewis Carroll's *Through the Looking Glass*, who said, "It's a pretty poor memory that only works in one direction!"?

Memory is inextricably linked to imagination. To improve your memory is to improve your creative imagination, and the reverse is also true. To remember something, it is important to link the new material to something familiar. By making a mental image of this new pairing, you are enhancing your chances of recall by wide margins. For example, when I was learning the Spanish word for frying pan (*sartén*), I visualized ten sardines with big red smiling lips stretching out of a skillet. The more novel or ridiculous the image, the easier it becomes to recall. This strategy can be used in all areas of learning, from discovering how the brain works to remembering the sequence of actions for throwing a bowling ball. The more you strengthen your memory with exercises, the better your creative skills become. Charlie Munger (partner of Warren Buffett) in his book *Poor Charlie's Almanack* says that this creative skill is critical for the near future as workers become compensated for thought processes versus physical labor. As you expose yourself to new situations and experiences, mentally connect these new stimuli to existing concepts. Bring unrelated concepts together to form new ideas. Children often join unrelated concepts as their imaginations run unchecked. When my grandson was barely four years old, he hugged his mother and told her, "Mommy, I love you like a nice hotel!" We found his simile hilarious, but to him, it made perfect sense. James Altucher and others call this process "idea sex." Your memory will improve, and you will become creative beyond your wildest expectations, for the thinking brain loves novelty.

Reason

Mary Morrissey calls reason our constant companion, and Bob Proctor finds reason a person's greatest gift. Eben Pagan describes how reason complements imagination: "In life, if the way that you think is better matched to what actually happens and the way the world actually works, it allows you to better imagine what could happen or what will happen. More precise thinking creates more precise analysis, estimation, prediction, explanation, decision-making, and much more."[15]

To illustrate how "reason" works, I'd like to share a story with you:

The Old Fisherman

A young fisherman traveled to a prime fishing spot he had heard about for years. After setting up and casting his lure, he was quite disappointed at the lack of biting fish. He noticed an older man downstream who was casting every few minutes. Upon closer observation, the young man discovered that the other fisherman was actually throwing several of his catch back in the water. "What's that about?" he wondered, walking closer to the lucky fisherman. As he approached, the young fellow asked, "Why are you throwing back so many large fish? Do they get bony as they grow?" The weathered old man looked at the young lad, then pointed to an old broken off ruler at his feet. "See that ruler, boy?" he asked impatiently. "I broke it off at exactly the size of my frying pan. I only keep the fish that will fit in my pan."

---

15 Pagan, Eben. *Opportunity*. p. 14.

While the fisherman's answer seems a bit humorous and foolish, it points to the limited thinking you and I might fall prey to by relying on unchallenged habits, our old frying pans.

Each of us has, at some time, used common reason to sabotage our efforts to move in a new direction by allowing our paradigms to pop up and remind us that something could go wrong or someone else must be to blame for an unpleasant situation.

On the other hand, when one employs *uncommon reason*, the tone changes to "Why not?" "Why not me?" and "Why not now?" By opening ourselves to the uncertainty of *uncommon reason*, we release ourselves from our self-imposed limitations.

My older daughter Kara applied uncommon reason to indulge her ambition to travel the world. Realizing she wasn't in her rightful spot in pursuit of a career, she secretly dropped out of college (just short of acquiring her business degree) and worked to save money. Despite many ensuing obstacles (losing her money reserve when her car broke down in a foreign country, having her travel girlfriend abort the trip shortly after they arrived in Europe, and having no viable plan for employment), she persevered on her journey, wanting to discover who she really was. Kara found employment in Germany, where she could work and travel in alternate time frames as the sports seasons changed. She worked as a chambermaid and a food server, and finally became a ski instructor. Imagine my surprise at that job, since she was a native Texan with little experience in the mountains! When I visited her in the Alps, her Swiss, German, and Prussian workmates (who had all skied before they walked) assured me she'd originally been hired on persistence, bravery, and "chutzpah" instead of skill on the slopes. They decided they could train someone to ski who was that courageous— so she was relegated to the back room to strap and fix skis, and eventually, became an excellent skier. On one of our rare phone calls (no cell phones then), Kara told me she'd secured a job as a wind surf instructor. I replied that I did not know she could wind surf. "I couldn't," she said, "but when the lake water is 50 degrees, you learn fast!" Her connection with other traveling workers led to almost a decade of employment in winter and summer seasons at a premium, world-renowned sports camp in Switzerland. She discovered her passion for teaching, and upon her return to the States, she began anew her college career on her own dime. After graduating, Kara became such an outstanding high school teacher that soon she was hired by her county education office to design courses to train other teachers.

This story is relevant to you because it illustrates that you *can* design and become your dream. Did Kara have difficulties and challenges along the way? Of course, she did. That's part of any huge life transformation. But who she became as she faced her challenges and what she learned about herself and about what inspires her is the priceless part of the journey. She later lived two years in Venice despite admonitions from Italian friends that doing so would be impossible. Kara now lives in a beautiful place that makes her heart sing, and she takes on causes that she never would have dared to tackle prior to her independence journey. Was it the orthodox manner her mother would have chosen for her? No, but it was the course that made her the brilliant woman she is today. If a shy young girl sitting in accounting class, wondering what she's doing there, can transform her courage and confidence, so can you!

Perception

Our HOTS have not been listed in any order of priority, for they work together, one playing a more prominent role, depending on the situation, and another taking the lead in a different circumstance. That being said, I must admit that perception is one of my preferred skills for rapid shifts. As I have seen in clients and students I have coached over the years in education, medical, and financial settings, the slightest change in a person's perception can open that individual for dramatically more positive results.

When my granddaughter left for her first day of kindergarten, she ran to the car, sporting her new backpack and hugging her lunch kit. The neighbor happily waved as she saw the youngster heading off to a new start. In the afternoon when the five-year-old returned, the neighbor, who was gathering her mail, eagerly asked, "Oli, how was your first day of school?" Dragging her backpack by one strap and holding her lunchbox low in two fingers, the young student, looking at her toes, answered glumly, "Twelve more years."

Clearly a shift in perception had taken its toll—and another shift would be necessary to change the course of her educational career!

Perception is the way a person views his or her own world. It is just that—a person's view of that moment in time. Naturally, our perception is influenced by our own experiences, by the beliefs we have formed, and by the beliefs and reactions of those around us. The point that most of us miss, however, is that the majority of those beliefs, impressions, and "meaning-making" assumptions are stored in our subconscious mind—out of view of our thinking mind. Therefore, we are often unaware of our blinders that limit our perspective. You and I cannot see the total picture when we're in the frame!

As you move forward to create a full-spectrum life that you love living, the road will become much smoother as you allow new ideas to expand your perception. Thomas Edison said, "If there is any one secret of success, it lies in the ability to get the other person's point of view and see things from that person's angle as well as from your own." We'll see in Chapter 8 that perception is the key tool for accomplishing the difficult and complex task of forgiveness.

In summary, the six Higher Order Thinking Skills we must recognize and put to work *for* us in an intentional manner are:

1. Imagination
2. Intuition
3. Will
4. Memory
5. Reason
6. Perception

Their application to your ultimate success in becoming the person living your ideal life will become apparent as you move through the steps in this book of transforming your life from where you are now to where you want to be. Knowing the names of these all-important mental faculties is less important than knowing their functions and recognizing them in action. To that end, let's look at a true story and see where the HOTS come into play.

When, at age ten, I became very ill with a sore throat and high fever, a spinal tap confirmed I had contracted polio. The next few weeks were a blur of searing pain, interrupted only by brief moments of relief from the steaming woolen strips placed with forceps on my legs and trunk and wrapped in heavy plastic. When the pain subsided about three weeks later, my legs were paralyzed and my entire right side was extremely weak. My attempts to move my legs were futile.

As my lung muscles declined, I was moved to the iron lung ward for breathing support. The other patients in the ward were encased in full-length iron lungs; I had a respirator that covered only my trunk (aptly called the turtle). My closest companion in the room (the only one I could see face-to-face, due to the position of all of our machines) was a bright-eyed blond boy named Kenny. Kenny had the widest, readiest, and brightest smile I'd ever seen, and his thick shock of golden hair with his twinkling eyes kept our stark little ward sunny. None of us spoke much, for breathing was an effort, and the loud moans of the iron lungs sucking in and out canceled most opportunities for chatting.

Each day my grandmother, who was in Europe when I became ill, sent me a picture postcard of the sights she was experiencing. I always had the nurse show them to Kenny, who seemed very interested.

One day, Kenny spoke hoarsely above the groans of the machines, asking me, "What does your grandma say to you?" "Oh, not much," I faltered, resisting the urge to share, for I had been severely censored by my mother and stepfather when I spoke of the lessons my grandmother had taught me as a young child.

Then, for some reason, I spoke openly about Grandma Clark's vision—that I was a child of God, made in the image and likeness of my Creator. She said I must "see" myself walking. I must not accept these circumstances as my reality; that was Error talking. I didn't have to understand *how*, but I did have to *know* I would walk. She said that image—already formed in her mind—would soon begin to form in the material world—from her own "knowing."

Kenny was silent for a while, and then he spoke with such clarity I hardly knew who was speaking. He said five words that changed my life:

"I can *see* you walking."

In that moment, something shifted within me. I *knew* I would walk. The seeds of Truth that my grandmother had planted earlier in my life suddenly bore fruit, and there was no longer doubt in my mind. In that simple sentence, I heard Kenny's faith. Through his belief, I was able to shift my perception and imagine a different result for my life. A still small voice inside me said, "This is Truth," and intuitively, I knew I would do whatever I needed to bring this reality into expression in the material world. Unconsciously, I was following the teaching of Wallace Wattles, one of the many authors my grandmother had read to me as a child. "Behind your vision must be the purpose to realize it."

That was the last conversation I had with Kenny. The following week, as the nurses wheeled out the huge machine he'd been confined in, we all knew his frail body had finally drawn its last breath. We each grieved the loss of his smile and the joy he gave us. I knew I would never forget him, for he held a belief so strong for my good that I could believe in his belief. Kenny, my earth angel, was only five years old.

Now let's unpack the HOTS in my story about Kenny. Where did imagination play a part? At first, only Kenny's imagination was working. When he clearly expressed his "image" of my walking, he empowered the sleeping imagination in me to wake up.

How about intuition—where did it come in? With Kenny's conviction so evident, his faith so real, I finally opened to the small voice telling me that indeed I would walk. This experience brings up a salient point about that intuitive voice: it does not always appear as a voice. Perhaps a billboard, or a book, or even a commercial on TV will grab your attention. Obviously, I had been receiving the same information from my grandmother, but I had been afraid to listen because my parents had denied and ridiculed this source of Spirit.

The skill of will—where did that figure in the scheme of things? It took calm, focused will to put my attention on the possibility and then probability of becoming ambulatory. I have a black-and-white photo of myself taken two years later. Our sixth grade class had a square dancing group, and I was able to join them. I secretly called this photo my victory shot, for it is similar to the image I held in mind for the twelve months I was unable to attend school because I was "crippled." I missed fifth grade entirely, but, except for the bulky support shoes I wore instead of the ballerina-like flats of the other girls, I fit right in with the other sixth graders!

How about memory—how was that important? In this case, it was necessary to "un-memorize" those many attempts when my limbs had failed me—and to "re-member" myself as a whole and healthy person made in the image and likeness of my Creator.

The use of reason came in when? Right from the outset, I needed to dispute common reason shared with me by many medical professionals who had declared my paralysis permanent. Uncommon reason was needed here to say that, although I couldn't conceive *how* it would happen, I had an inner knowing that it *was possible*, for I could clearly see my upright, walking body in my mind.

Perception was key. Can you see how? That's right. I had to shift my perception of myself as a weak, permanently disabled child to that of a strong human with the power to connect to all the love, joy, and goodness in the universe. Only by doing so could I heal my body (that already existed as perfectly healthy in the realm of infinite possibilities).

I hope these applications have helped edify the tremendous power of your HOTS. This system will support you in every manner you need as you become more aware of and adept at your use of these skills. Archimedes said, "Give me a place to stand and a lever long enough, and I will move the world." You have that lever—your HOTS. Use them wisely and joyfully. What glorious power you have to co-create a fulfilling life and in the process to become who you want to be! In the next chapter, you will create your perfect place to stand—your dream.

Throughout this book, we'll share many examples of transformations that have happened in spite of "impossible" circumstances. Clients, students, and patients of mine from all age groups and from different backgrounds have achieved spectacular dreams when they've lived the principle that one becomes her or his thoughts, after taking the necessary actions to manifest them. You will learn to marshal the amazing powers stored within you. You will understand how consciously to manage your HOTS (Higher Order Thinking Skills) as you apply the tested and proven principles in the art and science of becoming your dream.

"The greatest achievement was at first and for a time a dream. The oak sleeps in the acorn; the bird in the egg; and in the highest vision of the soul, a waking angel stirs. Dreams are the seedlings of realities."

— James Allen

CHAPTER 2

DESIGNING YOUR DREAM

HOTS ALERT: Imagination, Intuition, Memory

"Twenty years from now you will be more disappointed by the things you didn't do than by the ones you did. So throw off the bow lines. Sail away from the safe harbor. Catch the trade winds in your sails. Explore. Dream. Discover."

— Mark Twain

To transform your life, you must reach deep into your *imagination* to find what really makes you come alive. What are your secret desires? Discover and polish these gems as they come into your awareness. Respect those longings as Life wanting to express itself more freely and fully through you! Neville Goddard, a prolific metaphysical writer and influential spiritual thinker of the last century, wrote, "Everything we do unaccompanied by a change in our image is but futile readjustment of services," like rearranging the deck chairs on the *Titanic*. Throw off any constrictive thinking, disregard the practical notions that have kept you stuck in your current conditions, and fill your heart and mind with joy. Joseph Campbell, American mythologist and author, says, "The big question is whether you are going to be able to say a hearty yes to your adventure." In her book *You Are a Badass*, Jen Sincero teaches us that getting clear about our purpose "can be the difference between living a happy, fulfilled life of abundance, choice, and expansiveness or living in the restrictive veal pen of…indecision and tired old excuses." While I don't think we are necessarily born with a single purpose, as my purpose has shifted several times throughout my life, I fully agree with her assessment that when you find your calling for that particular life phase, "and you design your life in such a way that you can share your gifts with the world on a consistent basis, you feel like a rock star. When we share what we were brought here to give, we are in alignment with our highest, most powerful selves."[16]

As Benjamin Hardy points out in the blog, *Medium*, it is important to look to the future, not your past, while creating your dream. By looking forward to the future, we can feel gratitude for our past, but not be bound by its limitations. Reyna Aburto is quoted as saying, "Our path is not about what we have done or where we have been; it is about where we are going and what we are becoming." The founder of Strategic Coach, Dan Sullivan, teaches, "Always make your future bigger than your past…. We remain young to the degree that our ambitions are greater than our memories." Famous rock star Alice Cooper said about his creative process, "You have to believe your best work is ahead of you. If I believed my best song was already written, I wouldn't keep writing." John Burke, an Emmy-nominated pianist, says he always tries to do something he's never done before to push his creative boundaries. Richard Paul Evans, a thirty-eight-time *New York Times* bestselling author, echoes these sentiments, saying he tries to create something better than he's ever done each time he sits down to write a book.

Hardy informs us that our personality is simply an expression of our relationship with ourselves and what's around us. We can show up differently in the world, changing our personality completely. In fact, each new level of development will require an updated version of *you*. *Who* would you like to be, based on the future you want to have?

---

16 Sincero, Jen. *You Are a Badass.* p. 72.

You might *think* you have little imagination, but that's not true. To illustrate, let me remind you of the feeling you get when you're alone in a strange, dark place, and you hear unknown footsteps approaching. Your imagination will kick into gear immediately! So now let's use that wonderful mental attribute to your advantage.

As you consider the four domains of life listed below, I suggest you sit quietly and allow yourself to daydream fully. Let your mind wander until you can form the grandest image possible of yourself flourishing in that particular area. Dolly Parton says, "Find out who you are and do it on purpose."

Make this fun! This is *your* exciting life. Design it with all the zest you can muster! Be detailed and specific. In the words of John Henry Newman, "Fear not that your life shall come to an end, but rather that it shall never have a beginning."

Henry David Thoreau writes, "It is a gift to be able to paint a particular picture or to carve a statue, and so to make a few objects beautiful, but it is far more glorious to carve and paint the very atmosphere and medium through which we look. To affect the quality of the day—that is the highest of the arts." As you draw the images of your best self, living your best life, recognize that you are not only affecting the quality of *the day*, but you are creating the setting for *all the days* in your adventure here as a spiritual being having a precious human experience. Mary Morrissey advocates, "Build your want into a burning desire!"

Exercise

Let's begin with the area of **health and wellbeing.** We all want good health; include in this picture exactly what your vitality and vibrancy look like. What do you look like? What are you doing as a healthy individual? Are you outdoors? Where? Are you moving around? Who is with you? Are you taking deep breaths? What are you wearing? What sounds do you hear as you play?

_____

_____

_____

_____

_____

_____

_____

_____

Our next domain to complete is that of **relationships**. Include in this vision those relationships that are intimate, as well as friends, acquaintances, business relationships, and the community at large. Whom do you "hang" with? How do they respond to you? What do you notice about the characteristics of these different types of relationships? What are you doing that is making these relationships strong? The quality of your relationships—particularly with friends and those you love—is the single biggest correlate with longevity—so stick with this area a while. Define and picture these rich relationships in full color. Who just caught your eye? Bring that person in closer. Want to spend more time with him or her? Feel your heart and soul fill with excitement.

_____

_____

_____

_____

_____

_____

Now it's time to pay close attention to your **vocation**. Your vocation is how you spend your time and talents working, whether or not you are paid for your services. Where do you work? What do you do? With whom do you work, or are you alone? How long do you work each day? What does your work environment look like? How satisfying is your work? Do you have a service role? Are you valued in this role by yourself and by others? Do you look forward to this experience as you get up each day?

_____

_____

_____

_____

_____

_____

_____

_____

Finally, describe the **time and money freedom** you have in your ideal life. Do you have time to spend with your family and your friends? Are you free for leisure activities when they come up? For example, if you are a skier, do you have free time in the winter? Is there a passion you have always wanted to pursue, but time hasn't allowed you to build the necessary skills?

Money is useful for two things:

a. To meet your basic needs and keep you comfortable.

b. To allow you to give freely of yourself to those you love and to those causes that are important to you.

Do you have the cash that can fill these needs and allow you to do the things you love to do? What amount of money are you to earn in your dream life? Put an actual figure to this question after considering those things your heart desires to be, give, and do. The bank won't cash a check for "some money" or "enough money." Likewise, the universe needs specificity, so set an actual figure.

_____

_____

_____

_____

_____

_____

_____

_____

In *Opportunity*, Eben Pagan relates the philosophical idea of teleological causation—a fancy term that simply means a circular connection can exist between a goal we set and the cause. Telos is the ultimate end of something, such as an end state a person is moving toward. If the goal is important enough to us to go to work creating it, it may become the cause itself. That's how big and audacious I want you to make your dream life—something you can experience only by creating it. You have a friendly universe to support you in this transformation, so don't restrict yourself. You have the potential, the resources are available, and your opportunity is here now! The only thing missing is your decision to go for what you want.

As Pagan suggests, create a space of pure potentiality in your imagination, and then start inventing things for your future to put in that imagined space. What you need now is an aiming point, your place to stand, as Archimedes mentioned. When that is clear, we'll take the action steps into this uncharted territory. As Pritchett says in *You²*, when you jump, you don't think about the middle of the jump, but you focus on where you'll land. You'll be able to rely on the unseen forces we have discussed earlier, as imagination and intuition, and sometimes seemingly even luck, guide you to make the map as you go!

After you have formed a clear picture of your future life in each of these domains, it will be time for you to impress these visions on the universal mind. We live in a universal field of energy that is intelligent. We are also intelligent. These intelligences must come together to co-create the vision that Life is trying to express through us. By reading your vision each morning as you arise, and again at night before dropping off to sleep, you are demonstrating your intent and focus—your *will*—to have that which you desire. The wonderful news is that the universe, this great quantum field of energetic vibration, wants to give you your every wish even more than you want it! For a quantum leap to happen, you now must match the frequency of your dream. Low mental energy will not attract a higher vibration. We'll take the steps necessary to raise our own frequency as we go through the book, but first, you must test your dream to see if it is worthy of you. Is it a big enough dream, or do you need to revisit a domain or two to make them worth fighting for? Let's examine that in the next chapter.

"If the ladder is not leaning against the right wall,
every step we take just gets us to the wrong place faster."

— Stephen R. Covey

CHAPTER 3

TESTING, DIGESTING, AND INVESTING IN YOUR DREAM

HOTS Alert: Reason, Intuition

"Dream no small dreams for they have no
power to move the hearts of men."

— Johann Wolfgang Von Goethe

Test Your Dream

Now that you've designed your dream, let's test it to see if it's worthy of you. Most people ask whether they deserve their dream, but that's the wrong question. You'll spend your most precious commodity creating this dream—your time. As Maya Angelou says, "If you're always trying to be normal, you'll never know how amazing you can be." The first step toward making the invisible visible is to have an audacious, juicy goal to work for! You are not going for an incremental gain. You want a quantum leap! Fred Alan Wolf, in his book *Taking the Quantum Leap*, describes this term as follows:

> The explosive jump that a particle of matter undergoes in moving from one place to another…in a figurative sense, taking the quantum leap means taking a risk, going off into an uncharted territory with no guide to follow.[17]

---

17 Wolf, Fred Alan. *Taking the Quantum Leap*. p. 1.

You'll recall from our brief review of quantum mechanics in the last chapter that there are two types of energy: wave and particle. The wave is possibility, while the particle is reality. The third party involved is the observer. When observed, the wave collapses into a particle. By the very act of observing, the observer collapses the wave into a particle. This is scientific fact. So we, as observers and dreamers, must stop focusing on what is and focus on what we really *want*. The other point to remember is that particles make these "leaps" without covering all the bases, simply appearing in their new location. Therefore, the space and time dimensions by which we quantify our human experiences don't exist at the quantum level. It is necessary for us to reframe our concepts of ourselves and our universe and how we interrelate in order to use our magnificent creative powers. We are, at our basic level, energy, and we move about in a vast sea of energy. As we grasp this concept, we can begin to realize the significance of our own vibrational signature. More on this exciting topic, filled with new possibilities, later, but now it's time to use some proven tools to test our dream.

As a Life Mastery Coach trained by the *DreamBuilder* team, I often begin with Mary Morrissey's Five Questions to determine whether a dream is really life-worthy:

1. Does this dream make me feel alive?
2. Does the dream fit with my core values?
3. Is the dream going to require me to grow?
4. Do I need help from a Higher Power to accomplish this dream?
5. Is there some good in my dream for others?

If you can answer a resounding "Yes!" to each of these questions, you have a worthy dream. If not, your dream is not big enough.

"Get ruthless about trying something different," advocates Price Pritchett. Exercise your imagination more, and discover a dream that really lights you up.

Einstein said, "Imagination is everything. It's a preview of life's coming attractions." He also stated, "Logic will take you from A to

B. Imagination will take you everywhere."

Make the time and effort to be certain that your upcoming life is one you will love! Remember that you don't have a choice about creating. As long as you're breathing, you'll create—either by design or default, so design extravagantly!

Don Herold says, "Unhappiness is not knowing what we want and killing ourselves to get it." Make this dream one of greatness that brings you and those around you happiness and fulfillment.

Another of my mentors, Deepak Chopra, MD, offers a tool to reach deep inside and find your best dream. As you begin meditation or as you become still to reflect, ask these four questions:

1. Who am I?
2. What do I want?
3. How can I serve?
4. What am I thankful for?

Slipping into silence allows your true Self to bypass the mental chatter going on in your active mind so it can receive answers.

Vishen Lakhiani of Mindvalley, Inc. created what he calls "The Three Most Important Questions."

He recommends putting three columns on a sheet of paper and labeling them:

- Experiences
- Growth
- Contributions

Under each heading, write your heart's desires for your dream life. Many of us find this probe particularly helpful to "concretize" our vision.

My clients use all three tools to help dig past their current conditioned thought patterns and allow their true desires to emerge. I invite you to do the same.

Digest Your Dream

HOTS ALERT: Imagination, Perception

"Dream lofty dreams, and as you dream, so you shall become. Your vision is the promise of what you shall one day be. Your ideal is the prophecy of what you shall at last unveil."

— James Allen

You have fashioned a dream life that is explicitly yours; it stirs your passion and inspires your purpose. Recognize that each of us is expected by our Source to live an abundant life that serves us and our fellow man. You have tested your dream to be sure it is worthy of you. You constantly use your five senses to gather important information about the world you're in. Now bring your dream life into keen focus by using these senses of sight, sound, smell, touch, and taste to "cellularize" your vision. Make it palatable, real, and part of you! By doing so, your vibration will attract this dream from formless substance into material reality with quantum speed.

Thomas Edison said, "If parents pass enthusiasm along to their children, they will leave them an estate of incalculable value." Use his advice to fire up your *own* enthusiasm. Shift your imagination into the most daring, most exciting, and most outrageously fantastic life that will cause you to send waves of joy throughout the universe.

Stop reading now—and put your senses to work by adding such zest to each quadrant of your own dream life that your heart beats a bit faster and you anticipate goose bumps. Add all the emotion you can muster as you take your stroll through the images of your mind.

Make notes of the sensory input you can imagine as you crystallize the vision in your mind.

_____

_____

_____

_____

_____

_____

_____

Now it's time to add another dimension: your feelings.

How do you feel as you look at and touch your loved ones?

Are your sides splitting as you and your friends laugh, enjoying hilariously happy times together?

Who else is there? What connections have you reestablished in your free time?

What is it like to have all the necessary money, time, and creative ability to put yourself and your loved ones in your perfect home?

Where is this home?

How is it furnished?

Do you have a view?

What's the climate like? Can you feel the breeze?

Where do you exercise now that you're vital and healthy? Who's with you?

Where has your spiritual growth taken you? Do you have new awareness? A higher state of consciousness?

Are you enjoying your solitude for reflection and gratitude?

How has your passion materialized as your true purpose? Does that huge sense of fulfillment make you almost teary-eyed?

What about this very precious life you have created makes you want to shout out, "I *love* my life"?

As John Lovell puts it, "Ecstasy is a full, deep involvement in life." Ingest, digest, and immerse yourself in your dream.

Choose the *feelings* you want to experience as you become the person having your dream life.

How will you feel as a vital person of good health and wellbeing?

What feelings will you experience in each type of relationship? With your significant other? With your children? With your pet? With your extended family? With casual friends? With close friends? With colleagues in your work environment? With others in your workplace? With clients or customers?

What feelings will you have about yourself as you express your talents in the world?

Describe your feelings as you experience money freedom to be, do, have, and give what you want. Take some extra time, first as the giver, and then as the receiver.

How will you feel with freedom to use your time doing what you love and being with those you love?

At first glance, you might protest that your feelings will depend on circumstances. But, as a co-creator of this dream life, the truth is that you *choose* the feelings you will entertain.

Jot down some feelings that emerge.

_____

_____

Invest in Your Dream

HOTS ALERT: Reason, Will

"Don't ask what the world needs.
Ask what makes you come alive, and go do it.
Because what the world needs is people who have come alive."

— Howard Thurman

Before leaving this important step of testing your dream, let's look at another factor for success that you control: your commitment. While that factor seems an obvious statement, it is so important that it bears inspection. What determines commitment? Many people fervently wish for a different outcome, but they have little "skin in the game." The clearest indicators of your commitment are how you designate your time and money. Is this dream worthy of your total commitment?

Psychologist Benjamin Hardy has spent five years of doctoral research examining successful entrepreneurs and "wannabes." He found a clear distinction between the two types: Those people willing to go all in, to pass a "point of no return," are overwhelmingly the ones who succeed. These high achievers proactively create a point of no return by investing in the necessary training to enhance their skills and committing to the necessary time for practice until they reach mastery.

W. H. Murray, the Scottish mountaineer who was a member of the 1951 British reconnaissance team to Mount Everest, put it best:

> Until one is committed, there is hesitancy, the chance to draw back, always ineffectiveness. Concerning all acts of initiative creation, there is one elementary truth...that the moment one definitely commits oneself, then Providence moves too. All sorts of things occur to help one that would otherwise never have occurred. A whole stream of events issues from the decision, raising in one's favor all manner of incidents and meetings and material assistance which no man would have believed would come his way.

These leaders also arrange their environment to support their growth into new dimensions. Using information gathered from science, ancient wisdom, personal development leaders, and years of experience, I will share with you some of the details of others' successes. We will examine the strategies used by high performers so your quantum leap can become a reality!

Psychological, biological, and brain research available today indicate that anyone who wants to *can* change dramatically—*can* transform his or her life. This dramatic upgrade in awareness, consciousness, and skills does require commitment—usually of your time, money, and talents. All the Higher Order Thinking Skills must come into play to sustain your commitment throughout the journey.

Coach Jim Rohn says, "You should set goals that force you to become someone powerful in order to achieve them." Chapters 2 and 3 are designed to help you set those goals. Revisit these chapters often as you let go of old, limiting beliefs and raise the bar for your pursuits. At this point, the slogan of my coach Ryan Levesque is appropriate: "You don't have to get it perfect. Just get it going." Using your HOTS from Chapter 1, you will learn new strategies in Chapters 4–10 to become the person who fully lives your dream!

"Man often becomes what he believes himself to be. If I keep on saying to myself that I cannot do a certain thing, it is possible that I may end by really becoming incapable of doing it. On the contrary, if I have the belief that I can do it, I shall surely acquire the capacity to do it even if I may not have it at the beginning."

— Mahatma Gandhi

PART II

BELIEVING YOUR DREAM

Even if you have complete faith that you *can* achieve your dream, you might find Chapters 4 and 5 quite interesting because they speak to the insights that the many disciplines of science—e.g., microbiology, neuroscience, chemistry, and physics—are finally uncovering that illustrate the universal truths present since the beginning of life. I hope you will read carefully because this new information from scientists serves to deepen our understanding of the quantum laws that govern us (whether we know it or not). These laws give us such incredible power when we study and understand them.

Like many of us when we started this "life transformation business," you may have tucked away in your head a couple of "yes, but" responses, or a lack of faith that these miracles do actually apply to you. I understand because when I started studying high achievers several decades ago, I discovered a sneaky whisper in my ear. It kept saying things like "Yes, but she was an outstanding athlete," or "Yes, but he had the family and money backing that I don't have," or "Yes, but they are such a strong married couple that they can really support each other," or…or…or….

Finally, I caught on to my excuses when a simple example hit me in the face. A woman I knew started her own business in the '80s, when I first began my financial planning practice. Over the next few years, she enjoyed huge success, and I found myself—while admiring her and being happy for her—still thinking, "Yes, but she had a large network of people when she started." Suddenly, like a bolt of lightning, I recognized that, yes, she did have a large network. And why did she? Because of the groundbreaking work she had done prior to her "sudden" success. As I began to look deeper, I found the cold, hard, undisputable evidence that the high achievers *all*—without exception—had made choices in their important life decisions that put them in the path of future "overnight success." That point has now been illustrated in well-known books, such as *Outliers*, in which Malcolm Gladwell studied the lives of extremely successful people and found that their accomplishments came on the heels of many hours of practice to gain mastery. Even Bill Gates put in the hours on the computer to become great!

Your "yes, buts" may take a different form, but there's a good chance they linger in the dark recesses of your brain. For that reason, I have done my best to retrace the scientific points that erased my own doubts that these principles are universal, provable, and repeatable. I will share some of these with you in the following two chapters.

Part of the process of becoming your dream is an art form—the actual imagining of your unique life that life wants to express through you. But—and this is the big point of Chapters 4 and 5— part of this process is indisputable scientific fact. It's as real as the electric current running through your lines to so many points in your home—as real as the air waves transmitting information to all parts of the globe and beyond. No, I cannot explain it thoroughly because I do not understand the science of electric current or radio waves, but I do know they exist, just as you do. I use them every breathing moment in some way or another. Likewise, the only reason we—as whirling dervishes of energy—have doubts about our capacity to co-create our lives with the intelligent energetic field around us is that for centuries we have not had the correct information. Even though the myths controlling us have actually been disproved, old beliefs are slow to change. Many wars have been fought to maintain the status quo—for we, as humans, can find dramatic change—particularly in our established beliefs—very threatening.

Since you are embarking on this process of co-creation, I know you must be a seeker of knowledge and truth—particularly that which can propel you forward. In that light, I offer the next chapters. If you're already solid in your belief, in your faith in your own potential, you might skip to Chapter 6—where we begin the steps of movement. But please swing back to these two chapters after you finish the book to pick up new data and better understand the process of co-creation. Also, your own examples of the scientific and anecdotal data are always welcome! Just share them with all of us at JoanMcManus.com or BecomingYourDream.com.

"Rebels are the people who refuse the seen for the unseen."

— Anne Douglas Sedgwick

CHAPTER 4

BUILDING YOUR FAITH— KNOWING WHO AND WHERE YOU ARE

HOT Skills Alert: Imagination, Perception, Reason

"FAITH is the only agency through which the cosmic force of Infinite Intelligence can be harnessed and used."

— Napoleon Hill

After designing, testing, and adding richer flavor to your dream, you are ready to take the next critical step: *have faith in your dream.* This faith is not the same as a religious faith. It is the "I can" factor. It is believing you have the ability to bring your dream to pass as you follow the proven steps in this blueprint.

Great leaders have known the power of faith for thousands of years.

One of the most powerful demonstrations of faith mankind has seen was demonstrated by Mahatma Gandhi of India. Gandhi had no money, no warriors, and no materials of warfare. He had no home, not even a suit of clothes. Yet this amazing man believed completely that India's independence could be won if he persevered. He was eventually able to galvanize 200 million minds to move in unison as a single mind. Napoleon Hill in *Think and Grow Rich* reveals the source of his astounding power: "HE CREATED IT OUT OF HIS UNDERSTANDING OF THE PRINCIPLE OF FAITH. AND THROUGH HIS ABILITY TO TRANSPLANT THAT FAITH INTO THE MINDS OF 200 MILLION PEOPLE."

Martin Luther King, Jr. had such a strong belief, such faith in his dream of dignity for all people and of human rights for people of all races, religions, and beliefs, that he peacefully led thousands of people in diverse populations to make significant improvements in their working and living conditions. Even years after his death, Dr. King's faith that he *could* make a huge difference in human rights continues to inspire millions of us to act in this area.

Never underestimate either the need for or the power of *faith*!

Many of us have varying degrees of faith in our own ability, depending on outward circumstances and recent or past events. However, faith and commitment are necessary to be able to create a magnificent life, rich in all areas. Therefore, it's helpful to use some temporary aids to seed your faith until the actual unfolding of your dream brings such amazing evidence that you won't need these hacks to understand that you are operating with immutable universal laws. Here are three tips to aid you on your journey:

1. **Act as if you have complete faith.** If necessary, borrow some of my faith, or that of the many high achievers who have inspired you by passing boundaries dictated by common reason to attain greatness. Begin to perform each move as a person who knows the power of Universal Intelligence. As Price Pritchett puts it, you should act as if your success is a certainty.

2. **Keep an open mind.** Be open to the possibility that these principles will work brilliantly for you.

3. **Suspend your doubts.** Make an appointment with your doubts in three months to entertain their objections. By then, your evidence will send them packing! If you must doubt something,

doubt your own limitations. Meanwhile, adopt the "I can" mindset. When you change your mindset, you change your life's trajectory.

Take a moment now to be still and *know* that the dream is yours. By universal law, there is nothing you can conceive in your imagination that is beyond your capacity to achieve.

It is vitally important for you not only to believe in your dream, but also to believe in your ability to make it happen.

Each day, imagine yourself in your dream life. Become familiar with who you are as the person living the vision. What would you be doing, saying, thinking, and feeling if all the circumstances were already transformed? How convincing are you to the universe that you are already living your dream?

Remember, you want this change to happen in a quantum leap, not in the usual incremental steps. You are not "hoping" for a 10 percent increase in income; you are seeing exponential gains—gains that do not have to stop at all the bases—and gains that are not tied to our human concept of time and space. You now expect the universal laws that govern all of us to work for you just as surely as you expect the laws of electricity to work for you without knowing how they conduct the current.

If these concepts sound a bit mystical to you, stay with me as we move into some concrete examples.

The next step is to know *who* and *where* you are so you can actually understand the potential waiting for you right now.

Knowing *Who* You Are

> "You are not a human being having a spiritual experience.
> You are all spiritual beings having a human experience."
>
> — Pierre Teilhard de Chardin

Actually, you must "remember" who you are, for when you came into this earth journey, you were certain of your own magnificence. Just observe any toddler—she knows that the world is entirely her oyster. Only through social conditioning does a human forget her or his magnificence—and lose sight of the fact that she or he is created in the Creator's image and likeness.

Your essence is pure consciousness. Your future is created from the choices you make in each moment. The more you make choices from the position of conscious awareness, the more you will make spontaneous choices that benefit you—and those around you. Have you been making reactive choices instead of choosing proactive thoughts and actions?

List two reactive choices you made this week.

1.

2.

List two proactive choices you made this week.

1.

2.

In addition to being selective about your thoughts, it is important to recall that you are operating from a limited perspective. You cannot see the big picture. Always leave room for Universal Intelligence to "upgrade" your dream. Some of your desired outcomes might actually limit your ultimate goodness, so it is important to think and say, "This, or something better." For example, instead of limiting yourself to the outcome of one specific person's love and lifetime connection, you might ask that you are always surrounded by love—letting the infinite power of the Creator select from unlimited supply your perfect mate. The highest good of both parties is then served, and you get someone more wonderful for you than you could have dreamed. Or you could list all of the qualities you long for in a mate, and set about making sure you are the kind of person who will attract your ideal.

A balancing act is involved here. It is necessary to know specifically, in as much detail as possible, and with as much feeling as you can inject, *what* you desire, for that image activates your brain (reticular formation) to seek and scan for fitting opportunities. Yet, if you detach from the *exact* outcome while holding fast to the essence of your dream, this generous Universe (that I call God—the Grand Overall Designer) will grant your desires beyond your wildest expectations.

At this point, do not be concerned with *how* the change will occur. It is human nature to want to know the details of how anything will transpire. Change can be unsettling at times, for we crave to know how it will happen, how long it will take, and if the journey will be safe and pleasant. We can experience the thrill of spontaneity and wonder if we can let go of our need for the "how" details and open ourselves—even briefly—to uncertainty. This exhilaration is only possible for those who trust the universe has their backs. Helen Keller is my avatar for believing in the face of uncertainty. Her words offer me regular inspiration: "Security is mostly a superstition. It does not exist in nature, nor do the children of men as a whole experience it. Avoiding danger is no safer in the long run than outright exposure. Life is either a daring adventure or nothing."

This is the essence of faith—knowing that Spirit has your back. Faith, belief in your dream, and in your capacity to achieve your dream, is a necessary ingredient for transformation to occur. Often quoted is the scriptural analogy of needing faith only the size of a mustard seed to move mountains; you must, however, have that seed firmly planted in the fertile soil of your mind.

If your faith has weak spots at times, act *as if* it's strong. Borrow my solid belief that a rich, fulfilling life is your divine right. Soon your results in manifestation will replace your doubts and reinforce your belief in your power as a co-creator of your life.

Einstein said the most important decision one makes in life is whether or not this is a friendly universe. My early concept of a loving universe was shattered by the later discord of my childhood home life. I was too immature to recognize that my own thought patterns had shifted dramatically from my earlier happiness to bitter resentment, and that these internal causes were responsible for many of the effects I was experiencing. Before I could change other areas of my outer world, I would have to raise my consciousness in a more sustained manner.

However, at that time I *was* able to receive the affirmation of faith from young Kenny that, indeed, I would walk again. He spoke the truth with such clarity and at such an elevated, pure vibration that my doubt subsided. I did not think "maybe." I knew I would walk. I had no idea how the current circumstance of my total paralysis would change, but I knew it would. With renewed faith, I forged ahead. My recovery was not the delightful "skipping down the mountainside while drinking goat milk," that we read about in *Heidi*, but rather an arduous eighteen-month rigor of rebuilding muscles and rerouting nerve synapses. Although I'm sure I complained during the process, I did not doubt I'd eventually become ambulatory. In Chapter 5, I'll share why Kenny's positive thoughts could override my negative doubts.

This story is an example of being an "unconscious competent." I knew subconsciously how to respond mentally, due to my earlier training, but I did not understand that this change in thinking could be used in other domains of my life to completely change my relationships there as well. Hence, only the one area in which I held my faith transformed at that time.

My goal for you is to become a "conscious competent"—fully understanding and experiencing in real time the unfailing way these principles work. Only then can you put them to work in every domain of your life.

To have all the resources of Infinite Spirit available to you, you must build your faith so strongly that it can withstand the rough spots. You do that by understanding that powerful unseen forces are available to you right now—right where you are.

Quantum physics, said to be the most powerful science yet studied

by man, has forced us to reshape our thinking about our relationship with the universe. We must reconfigure our outdated concepts
of time and space to make use of the miraculous benefits available
to us. By doing so, by beginning to visualize accurately the compo-
sition of all things internal and external, you will understand not
only the possibility and probability, but also the certainty, of the
positive outcome of your co-creating with the Universe.

Understanding *Where* **You Are**

Our environment is a boundless sea of energy. It takes form from
the countless perspectives that mold their shapes onto the malleable "putty" of energy that fills every possible space. Hence, we can
see that the energy we emit attracts the results we find in our lives.
Wallace Wattles aptly named this substance "thinking stuff." His
description helps us remember that Divine Intelligence permeates
all matter and all space.

In addition to our moving about in a vast sea of responsive energy,
we can be further amazed to recognize that the subatomic units
we exist in have what is known as potentiality. Depending on the
thought and emotional waves that the observer (you, or I) projects
upon these units, they become waves or particles. Yes, that's right.
Subatomic energetic bits of potentiality only become waves or particles when they are being observed. Does that give new meaning
to "Notice what you are noticing?" If you dwell on your lack, you
bring more lack. If you consistently fill your thoughts and heart
with love, you vibrate out into the surrounding energy a higher
frequency that materializes as your highest desires.

These laws work just as precisely as gravity or any other physical
laws. Because you are a spiritual being having a human experience,
you have access to all the divine wisdom and intelligence you will
need to make profound changes and receive your dream life.

In the Western world, for more than 150 years, we have had an ob-
session with Newtonian physics, without regard to the subatomic

world of energy and matter. For example, Western medicine denied the mind-body connection in favor of a segmented approach to pathologies. What is now accepted as mainstream medicine was ardently disputed during the 1980s when I worked as a biofeedback therapist. Using electrodes connected to different points of the body, we therapists connected patients to machines that measured the electrical impulses the patients generated. The tone varied in pitch according to the amount of energy expressed by the nerves enervating the patients' muscles under the electrodes. With the feedback, participants quickly learned what they were "doing" body-wise and mind-wise to create particular frequencies. Indirectly, we were monitoring brain waves, for as the patient achieved deep relaxation, he or she moved from beta to alpha waves. Today, it is possible with advanced technology to measure directly the particular brain wave frequencies, giving the patient immediate feedback so that he or she can quickly duplicate the images and bodily sensations that produced the desired state.

Why is this work relevant to you now? It is a concrete demonstration of how our thoughts send electrical signals to the reptilian brain, which instantly communicates with the midbrain through tiny electrical impulses. The midbrain or mammalian brain (also called the limbic system) sends chemicals throughout our bodies as emotions that impact every cell in the body. Both of these sections of the brain operate together, usually outside our awareness. Together, they are considered our subconscious mind, which I call "Lizzie."

For simplification, I like to visualize that we have a three-brain system. The oldest, most primitive brain developed thousands of years before the midbrain evolved. This primitive reptilian brain, at the top of the spinal cord, controls autonomic physical body functions, such as heart rate, blood pressure, and digestion. The midbrain connects the forebrain and the hindbrain, and it controls emotional reactions to stimuli.

The last area to evolve was the prefrontal cortex (part of the neocortex), the thinking brain, which handles conceptual thought. This conscious thinking brain can manage all of the Higher Order Thinking Skills, making them accessible to our conscious control. Four of the HOTS—imagination, intuition, memory, and perception—actually reside in our subconscious mind. But, by using our unique ability to think about our own thoughts, we can override any self-defeating reactions from these mental faculties and reprogram them with empowering patterns to form new beliefs. The thinking brain can manage these skills for short bursts of energy so that they operate intentionally and in total awareness. Without regular monitoring, however, they fall prey to our habitual patterns of thinking that have formed well-worn paths in the subconscious.

The two HOTS that reside in the conscious prefrontal cortex, reason and will, also need regular inspection to ensure they are aligned with our highest purpose. For example, uncommon reason requires the positive emotion of courage, and must be accompanied by unfailing persistence to be effective.

New research shows that our prefrontal cortex can actually reframe some of our emotional reactions, which we'll discuss later. While the Higher Order Thinking Skills (HOTS) of imagination, intuition, memory, reason, will, and perception can operate intentionally, consciously, and in total awareness, they often fall prey to habitual patterns of thinking that have formed well-worn paths in the subconscious mind, so they must be checked-in on periodically by using our unique ability to think about our own thoughts.

Bruce Lipton, world-renowned microbiologist, describes in his outstanding book, *The Biology of Belief*, his laboratory discovery that the nucleus is not the brain of the cell. When the nucleus of a cell is removed, the cell continues to function! Further experiments revealed that the outer membrane of the cell has integral proteins that hook up with environmental signals to power the cell. The nucleus contains the DNA of the cell, but the membrane controls the function. As biologists—scientists who study living organisms—Lipton and colleagues earlier had myopically ignored the quantum world as unrelated to their field.

However, as quantum physicists probed the relationship between energy and matter, it became apparent that the universe is not made up of matter suspended in empty space—but of energy. Physical atoms are made up of vortices of energy spinning and vibrating constantly. Each emits its own energy signature. Atoms are made out of invisible energy, *not* tangible matter! Lipton quips that the emperor has no clothes. "So in our world, material substance (matter) appears out of thin air."[18]

Matter can be simultaneously defined as a solid particle and as an immaterial energy. (Hence the myriad of quantum mechanics jokes, i.e., "Schrodinger's cat went into the bar and didn't.") Remember the observer's influence from Dr. Quantum on YouTube? *Your environment responds to your vibratory signals.* You, in fact, are the affect that gives the effect to your conditions!

Einstein revealed that our universe is not made up of discreet physical objects surrounded by dead space; it is "one indivisible, dynamic whole in which energy and matter are so deeply entangled it is impossible to consider them as independent elements."

Because of our Western world bias for Newtonian physics, biomedicine has largely ignored the role that energy plays in health and disease. In the same manner, you and I have been trained through our education and culture that our external and internal circumstances are independent of each other. Instead of a linear flow of information, however, there is a continuous, multidirectional flow of information between subatomic particles and waves. This interrelational flow of energy renders our mind and body inseparable. Some scientists now refer to these structures as the mind-body.

---

18 Lipton, Bruce. *The Biology of Belief.* p. 71.

Information Flow

$$A \rightarrow B \rightarrow C \rightarrow D \rightarrow E$$

Newtonian - Linear

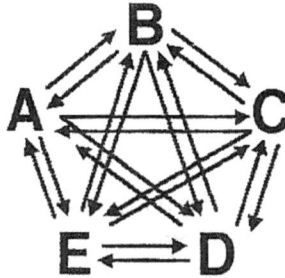

Quantum - Holistic

Multidirectional Information Flow Diagram[19]

Lipton describes another big "aha" that came later when he discovered that the mind's signals (a non-localized energy) override the localized body's signal. To prove his point, the biologist gives many examples of controlled studies involving the "placebo effect."

He further describes the advantages of our dual mind—the conscious and subconscious mind. The "thinking self" nature of the mind he likens to "a ghost in the machine," while the subconscious mind is more like a jukebox programmed to play a certain playlist. There is no point in reasoning or battling with the subconscious mind. After a while, it will follow its own tape player. Only by changing the behavioral tapes can we override the program. (A clear illustration of why "willpower"—the struggle—does not yield long-term success in pattern changes.) Furthermore, the tension between our conscious willpower and our subconscious mind can, and often does, result in serious neurological disorders.

---

19 Lipton, Bruce. *The Biology of Belief.* p. 73.

Do you see where I'm going with this? The implications for us are staggering! You and I have been designed to create our external circumstances by carefully and selectively choosing in our conscious thinking mind the information we feed or allow into our subconscious automatic mind. We have been given the gift to choose what our focused attention will notice and engage with. Due to brain structures, such as the reticular formation, our eyes will only see and our ears will only hear what we are selectively noticing.

As Bruce Lipton was doing his work with the membrane-based cellular functions, Candace Pert, PhD, was searching for the biochemical links between consciousness, mind, and body. In her book, *Molecules of Emotion*, she describes her journey for truth, discovering that neuropeptides and receptors work as information molecules. These bio-chemicals of emotion are the informational content that is exchanged—via the mind-body network—with the many systems, organs, and cells participating in the process. "Like information, then, the emotions travel between the two realms of mind and body, as the peptides and their receptors in the physical realm, and as the feelings we experience and call emotions in the nonmaterial realm."[20]

In my experience, emotions sometimes precede a formed thought. Haven't you experienced a rush of emotion, such as love, jealousy, or anger, that actually surged through your body before your mind caught up? I mention this phenomenon because the energetic vibration of any word or thought is certainly dependent on the emotional charge that accompanies it. To view our thought-words without simultaneously analyzing the accompanying emotion is to receive only part of the data.

Pert calls information the missing piece that transcends the Cartesian view of the mind-body split, for information belongs to neither mind nor body, but touches both. She describes the new self-image as "one of an integrated body and mind, one with intelligence, an emotional intelligence, even a soul or spiritual component. And the undeniable implication [is] that each of us is a dynamic system with a constant potential for change."[21]

20 Pert, Candace. *Molecules of Emotion*. p. 261.
21 Ibid. p. 262.

Our emotions are basically physical reactions we have when our brain, after interpreting a triggering event, sends electro-chemical signals to every cell in our body in response to some external or internal stimulus. Your default pattern when you hear a negative tone of voice or feel a surge of fear is the unconscious choice you have repeatedly made when experiencing a similar stimulus.

Emotions, which are generally instinctive, generate an emotional response like fear, sadness, anger, relief, amusement, or love. Feelings, as described here, are mental interpretations of the emotion. When such an impulse is generated, many people simply accept whatever time-worn brain habits they have developed. Thus emotions and their responses remain buried at a subconscious level, whether or not they are life-supportive or degenerative.

This is not your only option! The *big truth* is you do not need to continue allowing whatever emotion makes an appearance to follow that old, non-productive route. Instead, you can make a tactical choice about how you intend to respond. Brendon Burchard, internationally recognized performance coach, states this concept clearly in *High Performance Habits*: "Think of an emotion as mostly a reaction, and feeling as an interpretation."[22]

Burchard details how an Olympic athlete he coaches describes this phenomenon. "When I'm at the starting blocks, my body is naturally aware of what's at stake, and there's an emotion of [some fear] that's there no matter what. But I don't feel anxious. I define the feeling. I tell myself that what I'm sensing is a feeling of readiness, excitement."[23] With thousands of subjects, Burchard found that high performers have the emotional intelligence and will to anticipate their predominant reactions to situations and design their own meaning of that emotional response when it occurs.

These are compelling reasons to spend more time with your dream, anticipating your emotions and *intending* the feelings you want to incorporate into the situation! Go through the questions above.

---

22 Burchard, Brendon. *High Performance Habits*. p. 80.
23 Ibid. p. 79.

Armed with this new information, you can actually determine how to frame the emotions that come up. Become more aware of those subtle little pulses of energy that indicate your midbrain has sent an electrical charge and chemicals called peptides (an electromagnetic force) to all parts of your body. Decide what feeling(s) you want to bring to an interaction. Then decide what feeling(s) you want to get from the situation. Set your intention for the feelings you want to endure. Constantly work to build a support system—a framework—for the responses you will incorporate in all your interactions with others and in your own self-communication. This option is part of the miraculous life we are living as the only species able to *choose* the feelings we attach to different situations!

Both Lipton and Pert give scientific proof that the reductionist theory that Western medicine has preferred since the sixteenth and seventeenth centuries no longer works as our model of the universe. Instead, relatedness, cooperation, interdependence, and synergy have replaced simple force and response. The immensely complex psychosomatic (mind-body) network with trillions of shared components—the receptors and neuropeptides—works throughout our many systems and organs.

When I mentioned to Dr. Deepak Chopra, who was teaching a group of us, that I was pleased science is now validating the truths from ancient wisdoms, he corrected me by saying I had it backwards. Science is finally catching up to these truths contained in spiritual laws. (Had I had complete faith in the universal laws, I would not have needed to look outside myself for validation.) We now have thousands of controlled studies of both the microscopic and subatomic levels of matter and energy that demonstrate the truth for those of us who still choose to be skeptics.

For the reader who would like a more in-depth understanding of these spectacular findings, I highly recommend the two books mentioned above: *The Biology of Belief* by Bruce Lipton, PhD and *Molecules of Emotion* by Candace Pert, PhD. Each—although quite detailed—is intended for the lay person and will shift your perspective on the reality of the power of you, the blessed being, made of vibratory magnetic energy, having potentiality—an infinite number of possibilities—for the kind of precious life you choose to live. As you increase your understanding, your faith will become boundless!

Dr. David Hawkins, MD, PhD, in his excellent book, *Power vs. Force*, gives more evidence of the profound effect of vibratory frequencies in our lives and in our behavior. In 1971, Dr. George Goodheart established what he called "applied kinesiology." He determined that muscle response was correlated with harmful or beneficial stimuli. If the deltoid muscle (usually the one tested) instantly went weak, the stimulus was harmful to the body. If the extended arm remained strong, the substance was therapeutic to the body. Thousands of practitioners began to use this method for reliable diagnostic information about a patient's individual response to treatment. Used extensively by holistic practitioners, it did not catch on in mainstream medicine. Researchers used kinesiology as a useful technique in determining allergies, nutritional disorders, and responses to prescribed medications.

While kinesiology was accepted widely in many circles, until recently, the medical profession in general, and psychiatrists in particular, soundly rejected the proposition that such a noninvasive process could be effective and valid. They were hostile to the 1973 book *Orthomolecular Psychiatry* by David Hawkins, MD, PhD, and Nobelist Linus Pauling, which illustrates how beliefs can thwart open-mindedness to new information. The book's thesis was that mental illness, such as psychosis, as well as emotional disorders, could be affected by nutrition as well as medication.

Dr. John Diamond, a psychiatrist, used this kinesiology in diagnosing and treating psychiatric patients, labeling the method "behavioral kinesiology."

He refined testing techniques so that over the last two decades, practitioners and researchers have universally observed that test responses are completely independent of the subjects' belief systems, intellectual opinions, logic, or reason.[24]

---

24 Hawkins, David, and Pauling, Linus. *Orthomolecular Psychiatry*. p. 60.

Dr. Diamond researched more psychological stimuli, such as music, art forms, facial expressions, voice tones, and emotional stress. His work excited many others to pursue modes of applicability. The findings showed that the body responds accurately—even when the conscious mind is naïve! Some physicians, including Dr. Hawkins, have entered the lecture circuit to demonstrate this amazing body intelligence.

The standardization of calibrating the vibratory frequency of each distinct emotion is where this information becomes relevant to each of us. To know the actual vibratory signal we are emitting when experiencing particular emotional states is a strong motivator to assume leadership of our mind!

The chart below indicates familiar emotions as a reference point. The frequencies below 200 are considered to be harmful to life, while those from 200—1,000 are beneficial. Normal human range is 1—600, while the frequencies from 601—1,000 are only seen rarely in spiritually advanced, enlightened people.

The frequency of LOVE is 500; therefore, I use the mantra of "Love and above" when raising my mental state. By focusing on a person for whom I feel unconditional love, I can shift my thoughts out of petty occupations with lower frequencies. For example, I used to struggle with jealousy. I was happy when I observed someone's success, but I also entertained a throat-tightening feeling of envy, with an immediate reaction of "Wish I…." By letting go of that thought and moving to a firmly established "icon" of love in my mental storage, I could shift from the often negative vibes I was emitting. I offer this strategy as a technique to employ when your intelligent body sends a signal that you've regressed—even below the 200 level of integrity into the harmful, negative, life-sucking emotions. Your negative emotion may be at the level of fear, anger, or another level, but the same process can relieve you and rewire those automatic responses that the subconscious midbrain and cerebellum have diligently programmed in you to "keep you safe."

Level	Emotion	Log
Enlightenment	Ineffable	700—1,000
Peace	Bliss	600
Joy	Serenity	540
Love	Reverence	500
Reason	Understanding	400
Acceptance	Forgiveness	350
Willingness	Optimism	310
Neutrality	Trust	250
Courage	Affirmation	200
Pride	Scorn	175
Anger	Hate	150
Desire	Craving	125
Fear	Anxiety	100
Grief	Regret	75
Apathy	Despair	50
Guilt	Blame	30
Shame	Humiliation	20

The work of Pert and Lipton confirms that our minds and bodies have signature vibratory patterns depending on our thoughts and emotions. This energy does not stop at one's body, but extends to the environment. Have you ever entered a room where people have been arguing? Remember the uncomfortable "vibes" that remained, whether or not the dispute was still in progress?

Following is a story that illustrates the power of vibes.

The Gatekeeper

A traveler approached a great, walled city. Before entering its gates, he stopped to talk with an old man seated beneath a tree.

"What are the people like in this city?" asked the traveler.

"How were the people from where you came?" queried the old man.

"A terrible lot," grumbled the traveler. "Mean, miserable, and detestable in all respects."

"You will find them here the same," responded the old man.

A second traveler soon happened by. He, too, was on his way to the great city and stopped to ask the old man about the people he would soon meet there.

The old man repeated the question he had asked the first traveler. "How were the people from where you came?"

To this the second traveler answered, "They were fine people. Generous, kind, compassionate."

"You will find them here the same," observed the old man.

This oft-repeated story reflects the truth that our dominant thought patterns create our results. We think not based on what we see, hear, or experience with our five senses, but from a different level of our being. Therefore, we should review our thinking—notice what we are noticing. In *Working with the Law*, Raymond Holliwell states it clearly: "It is evident, therefore, that of all of the factors which regulate the life and experience of the person, none perhaps exercise a greater influence than the ruling state of mind."[25]

We are unique creatures in the universe because we are the only ones (as far as we know) who can control our thoughts by choosing where to place our attention. This is not a gift awarded for some fantastic accom-

25 Holliwell, Raymond. *Working with the Law*. p. 28.

plishment of ours; it is given to us as offspring of the Creator. Because we are living, breathing humans, we have the choice whether to live our lives by design or by default.

Never doubt your worthiness. You have been given a priceless gift of choice—the choice of how you are going to live this precious life of yours. Don't accept only the low-hanging fruit. Go for the perfect prize to fulfill your deepest desires! As Raymond Barker states in his book, *The Power of Decision*, "None of us is intended to be average."[26]

It is my firm belief that you are a genius in some field of living.

Mt. Everest climber Edmund Hillary said, "People do not decide to become extraordinary. They decide to accomplish extraordinary things." He said further, "You don't have to be a fantastic hero to do certain things—to compete. You can be an ordinary chap, sufficiently motivated to reach challenging goals."

Too many people give up seeking a larger life, relying only on the evidence of their five senses to determine what is possible for them, instead of harnessing the power of their invisible mental faculties, their HOTS. By adding faith—the belief that you can have what you desire, and that you can accomplish the transformation by following proven and replicable steps—your success is established.

To clarify and reiterate the important ideas in this chapter, I will borrow from Wallace Wattles, who writes in *The Science of Getting Rich*:

> There is a thinking stuff from which all things are made, and which, in its original state, permeates, penetrates, and fills the interspaces of the universe.
>
> A thought, in this substance, produces the thing that is imaged by the thought.
>
> Man can form things in his thought, and, by impressing his thought upon formless substance, can cause the thing he thinks about to be created.[27]

---

26 Barker, Raymond. *The Power of Decision*. p. 160.
27 Wattles, Wallace. *The Science of Getting Rich*. p. 15.

The basic fact behind all appearances is "There is one Thinking Substance from which and by which all things are made."

I'll use the metaphor of making cookies to bring home these concepts. Years ago, in Germany, I purchased an antique set of tin cookie-cutters. Intricately formed shapes were bounded on the top by a flat edge, but on the bottom, the edges were quite sharp. When I placed the cookie-cutter (which in this analogy represents thoughts, images, and words) on top of the dough (which represents the external environmental sea of energy), nothing much happened. But when I pressed firmly down into the dough, I cut a delightful shape. This pressure represents emotion bound with faith. Just as the baker can select the exact desired cookie shape by choosing the preferred cutter, so you can form the mental image of your desires. By adding strong feelings and unfailing faith to the process (you certainly would not doubt that the sharply edged cutter could slice through the soft dough), you can be certain that your dream—already formed in your mind—is on its way to materialization.

"Sow a thought, reap an action; sow an action, reap a habit; sow a habit, reap a character; sow a character, reap a destiny."

— Stephen R. Covey

CHAPTER 5

UNDERSTANDING THE POWER OF YOUR THOUGHTS AND WORDS

HOTS ALERT: Imagination, Memory

"All progress, all success springs from thinking."

— Thomas Alva Edison

The Power of Thought

In the last chapter, we looked at who we are and where we exist. By understanding that we are energetic beings living in a vast energetic field that has the potentiality to transmute into material form, we can recognize not only the possibility, but also the predictability of our power to co-create our lives. This line of thought is by no means new to our country. The New Thought Movement, the first truly North American philosophy, evolved during the late nineteenth century. Philosophers, businessmen, and thinkers began writing about the power of positive thinking and positive thought processes based on a secular point of view. Through poems, essays, and nonfiction writing, men and women contributed to the movement, creating their own works and drawing on those of earlier powerful thinkers like Ralph Waldo Emerson, Henry David Thoreau, and Louisa May Alcott. Ernest Holmes, regarded as the Father of Positive Thinking, stated that thoughts are things. Philosopher William James and Pharmacist Emile Coue also stressed the importance of one's thinking. James reiterated in many books: "The greatest discovery of my generation is that a man can alter his life simply by altering his state of mind." Emile Coue advanced auto-suggestion, creating an affirmation that I use regularly: "Every day in every way I am getting better and better."

These principles, combined with the phenomena described by the energetic subatomic world, explain the earlier statement that we are always creating, either by design—carefully choosing the thoughts that receive our attention and the words with which we describe our experiences—or by default, carelessly allowing our random, habitual thoughts and words to come to fruition as "more of the same" that we are declaring. We don't have the option to "not create." Although it takes effort, persistence, and continuity, we can change our thoughts and words to become forces for rapid positive growth.

Thomas Edison reinforced this idea by saying the "brain can be developed just the same as the muscles can be developed, if one will only take the pains to train the mind to think. Why do so many men never amount to anything? Because they don't think." For transformational change to occur, there must be a radical shift in our attention. We must choose our thoughts and words to conform to our new reality.

This point implies awareness, which brings us to the Number One Rule in upgrading your thoughts:

Notice what you are noticing.

In other words, become consistent in checking your thoughts. Use your valuable metacognitive skills (meta = above, cognition = thinking; thinking about your thinking) to determine your dominant mental patterns. This practice involves using your highest and most powerful intelligence. Even if it's been latent, you can strengthen the ability through awareness and practice.

Checking Your Dominant Thought Patterns

You can easily check out the positive or negative charge of your frequent thought patterns by looking at your current conditions and circumstances. What you focus on mentally will manifest in your material world. As Neville Goddard put it so well in *The Power of Awareness*, "Inner talking matures in the dark. From the dark it issues into the light. The right inner speech is the speech that would be yours were you to realize your ideal. In other words, it is the speech of fulfilled desire. 'I am that.'"[28] Goddard expands this idea as follows:

> Inner talking reveals the activities of imagination, activities which are the causes of the circumstances of life. As a rule man is totally unaware of his inner talking and therefore sees himself not as the cause but the victim of circumstance. To consciously create circumstance man must consciously direct his inner speech, matching 'the still small voice' to his fulfilled desires.[29]

This alignment is crucial, for "inner speech is always objectifying itself."[30]

I originally found this spiritual and scientific truth to be tough to swallow. I don't know how you feel about your thinking, but since you are a student of self-advancement, I surmise that you (as I did) think you have a predominance of positive thought patterns, as well as sound beliefs and values that shape your thoughts. I thought my intentions to evolve and expand into an ever-improving version of myself were clearly imbedded in my subconscious.

I was surprised, however, to discover that worry, anxiety, doubt, indecision, anger, impatience, and irritation played a more prominent role in my internal monologue than I realized. As Norman Vincent Peale taught many years ago: "What the mind dwells on, expands."

---

28 Goddard, Neville. *The Power of Awareness.* p. 163.
29 Ibid. p. 166.
30 Ibid. p. 170.

When I began seriously taking inventory, what I observed was a surprise! Many moments of negative instances added up to some general biases, judgments, and cloudy visions that I did not know I harbored. Yet they were working forcefully in my subconscious, causing my ever-vigilant "lizard brain" and midbrain (our ancient limbic system) to reach out and sabotage the changes I wanted to make.

You see, that dear old reptilian brain, working intimately with the younger but still ancient mammalian brain ("Lizzie," as I call them) has been entrusted for thousands of years with our livelihood—not just our occupation—but our actual staying alive. It faithfully thwarts any plans our conscious prefrontal cortex makes to put us on a different course, because the unfamiliar—in caveman days—was often fatal. Even though we no longer regularly face saber-toothed tigers and other elements of nature that could wipe us out in a blink, our primitive cerebellum, the seat of the subconscious, still works to protect us at all costs, engaging all parts of the primitive brain.

Listening carefully to any and all sensory data input from the environment, our thoughts, and our experiences regarding money, health, others' motives, the bazillion "shoulds" from well-meaning folks, and the onslaught of sensational news from the media, our subconscious mind remains constantly alert to detect any matches with previous memories that involved a threat. This loyal "servant" becomes the master, working twenty-four hours a day to excite and ignite those hard-wired cautions we've accrued through our years on earth! Simultaneously, the midbrain sends chemicals to all parts of the body. This emotional center communicates chemically with every cell in the body through peptides, making the physical body react to the subconscious impulses. Lizzie means well, but she doesn't know the whole picture because she cannot reason. She simply repeats her playlist like the jukebox she is. It is our job to rewire and refire new positive beliefs and patterns for the old gal to work from.

After many years of "cleaning out" the debris, I still find myself encountering erroneous beliefs and inefficient patterns that must be replaced. However, it has become much easier to detect thoughts heading in a non-productive, non-loving direction. I have learned, through my studies and coaching, to apply several tools to correct and redirect my thoughts. I'll share these with you as we examine challenges in upcoming chapters.

Again, what is Rule #1? Mary Morrissey calls this the #1 rule of Brave Thinkers:

Notice what you are noticing.

Why? You have the divine potential to create whatever you put your attention on. More than that—you *will* create what you put your focused and sustained attention on!

There is a caveat to this good news. The creating part is non-negotiable. You are continually creating your results in all areas of life—whether you are aware of your choices or not. If you try to avoid creating by staying in bed with the covers over your head, what you have created is a day in bed with covers overhead.

Here we can recognize the magnitude of this revelation of your power. What you focus on sets the thought vibration of your dominant mind patterns. If you put your attention on the current circumstances that trouble, frighten, or annoy you, you will reflect the frequency of those thoughts in your external results.

If, however, you keep your attention on your *own* dream, not the dream of your culture, your family, the media, or society at large (which Vishen Lakhiani so aptly named your culturescape), your mental frequency will reflect this higher vibration, attracting what you love. This law—the Law of Attraction—is not psychology—it is physics. Despite some media distortions of the Law of Attraction, and despite the missing pieces in some explanations of the law, it is scientific fact—a subset of the Law of Vibration.

To become the person you were intended to be, however, you must also "decide" to eliminate false thoughts of limitation and give up your accrued negative habits. (To "de-cide" literally means "to cut away.") Bringing this idea back to the cookie dough metaphor, after the baker presses the chosen cookie cutter with fine, sharp definition into the dough, he makes a decisive twist of the wrist before bringing up the cutter with dough in it. This sharp cutting action separates the excess dough from the desired cookie design, allowing unneeded pieces to fall away. In the same way, the following chapters will teach "the twist" to purge beliefs and habits that are no longer effective or efficient—that no longer fit the details of your dream.

Take some time now to examine some outdated beliefs you carry.

- Look at any of your non-acceptance of differences in others that have caused you to feel separated from the rest of humanity.

- Detect those "slips of tongue" or "slips of mind" that betray alignment of thought with your vision.

- Inspect and replace beliefs passed on to you by parents or authority figures that you can now recognize as the falsehoods promoted by fear or by the urge for control.

- Examine your thoughts for images of yourself as separate from others or from your environment.

How do these beliefs influence your idle thoughts? How do they influence the casual language you use?

Can you begin to see an actual physical advantage to your body when you forgive others or yourself or release resentment? (We will discuss this significant topic of forgiveness in depth in Chapter 8.)

If you can conceive and believe it, you can achieve it. You are capable of exponential improvement in your performance. You have barely scratched the surface of your potential. To change that fact, you must recognize and employ the power of your thoughts and words. You *can* think strategically and with intention. You have been given the supreme gift of choice about your thoughts. This choice allows you to—by sustained selected thought—become crystal clear about what you want and then to become open to the action steps needed to bring about your transformation.

Many great thinkers, such as Thomas Edison, have advocated that the best thinking is usually done in solitude, while the worst is done in the turmoil of daily life. Edison quipped that man will go to great lengths to avoid thinking. Therefore, to make our environment optimal for intelligent use of all our HOTS, we will set up a critical morning routine in the next chapter.

You cannot remain at your highest level of thinking if you are surrounded only by naysayers, victims, jealous people, or those without hope or faith in their own abilities. For that reason, it is important to select one or two (or more) people with whom you can share your dream in complete confidence that they have your best interest at heart. They must share your faith in the principles we have discussed—that you are co-creating with Infinite Intelligence a life of purpose, fulfillment, and happiness. Without this support, maintaining a high vibratory level of thought infused with loving emotion becomes much more difficult, for the gravitational pull back to your old patterns is unbelievably strong. In *The Power of Decision*, Raymond Charles Barker equates placing your attention on positive people with affirmative prayer because it "feeds into your subconscious mind the mental materials which then garner emotional equivalents to bring about the objects of your attention. It is the success process, the success mechanism."[31]

Barker goes on to say that when you are worried, you are using this process in reverse, creating a mental and emotional movement "toward the evil end of the stick."[32] Barker emphasizes that controlling where you put your attention requires self-discipline. It's not easy, and it must be motivated by desire. When you do accomplish this focus, however, it puts you at the level of solution instead of the level of problem. We will cover steps to find a suitable mastermind to reinforce your "right thinking" in Chapter 8.

I speak with certainty about our ability as thinking humans to shape our results in every domain of life because I have experienced these principles in action in my own life, and I have witnessed their successful use by hundreds of clients and students

---

31 Barker, Raymond Charles. *The Power of Decision*. p. 98.
32 Ibid. p. 98.

who became aware of their mental focus and changed it to reflect the circumstances they desired instead of materializing their fears. As an educator and coach, I fully concur with Henry Ford, who said, "The object of education, as I see it, is not to fill a man's mind with facts; it is to teach him how to use his mind in thinking."

As you continue your journey harnessing all of your HOTS—your *imagination, intuition, reason, memory,* and *perspective*—as you become facile with recognizing these mental powers and using them to your advantage, you will know how valuable "right thinking" (to use the term coined by Wallace Wattles) is to your complete success. Henry Ford assured us, "There isn't a person anywhere that isn't capable of doing more than he thinks he can."

List three doubts about your own capabilities that you have discovered as you peeked into your inner thoughts.

1.

2.

3.

What are some positive affirmations you can use to replace those erroneous thoughts as they try to take hold?

1.

2.

3.

The Power of the Word

> "Without knowing the force of words, it is
> impossible to know more."
>
> — Confucius

Most religions and philosophies have taught that the word is powerful, yet we humans have come to disregard the significance of this truth.

However, it becomes easy to understand the importance of our word choice when we consider the premises described in Chapter 3:

- There is everywhere a "thinking" substance.
- This substance filling all space is malleable to thought/word impressions.
- Man can choose which thoughts and words are to become his reality.

Thus, if one persists in complaining about circumstances that are not optimal, or continues negatively judging others, or feels hopeless about the future, these unfortunate constructs only reinforce and multiply what already is.

Thoughts extend beyond our bodies into the space around us. Think, then, how strong the force is of words that are true vibrations projected purposely outward!

A series of amazing scientific studies have been performed by internationally acclaimed researcher Masaru Emoto. His work with water crystals left no doubt about the invisible but profound effects of different words. In brief, when positive words written on a card, such as *love, peace,* and *joy,* were taped to a vial of water and the water was frozen, the magnified ice crystals formed exquisitely beautiful designs. By contrast, when negative words (*hate, anger, despair*) were attached to the vials, the resulting enlarged crystal designs were ugly, erratic forms. Emoto went on to discover that music, also a vibration, created corresponding shapes depending on the type of music. He learned that great healing power for humans comes from blending the sounds of carefully chosen classical music (Debussy, Strauss, Ivanovici, and Smetana) with photographs of these elegant water crystals. Why is this so important? Because we can readily see the undeniable impact of different types of vibrations on energy and matter. Knowing that the "thinking stuff"—also called consciousness—can be shaped and molded by the vibratory energy of our word-thoughts gives us the freedom to become our dream.

A 2007 Harvard study by Ellen Langer and Alia Crum tested the hypothesis that verbal information would actually change the physical outcome of exercise. From seven hotels in Boston, the researchers studied eighty-four housekeepers, who cleaned an average of fifteen rooms a day. Each room accounted for about thirty minutes of exercise. They informed the maids from four of the hotels that their exercise met the Surgeon General's requirement for healthy fitness. They did not inform the maids from the remaining three hotels about the value of their exercise. After just four weeks, the informed group had lost an average of two pounds each, reduced their blood pressure by 10 percent, experienced a reduction in their body-to-fat ratio, and had a lower BMI (body mass index). The uninformed group showed no change in any of these measures!

An earlier Harvard study in 1996 by Alvaro Pascual-Leone and his colleagues involved students who learned a simple five-finger exercise on the piano. Half the group practiced twenty reps each day on the piano. The other half only imagined practicing the twenty reps each day. The before and after mappings of their cortical regions responsible for moving their fingers gave some astounding results. There was a 29 percent increase in the brain structure of the students who actually practiced on the piano, and a 23 percent increase in those same areas in the brains of the students who only visualized their practice! You *can* develop new habits and change the structures of your brain. You *can also* develop your mindset and change your brain structures using your power of thought and words!

The same principle can work against us. As Carl Jung pointed out,

"Until you make the unconscious conscious, it will direct your life, and you will call it fate." Those buried thoughts that have become stories in our subconscious must be examined. Are they beneficial or destructive to our vision? We must be continually alert to our interpretations of past experiences. Those stories can be changed, for our HOTS' memory is malleable. What really happened is insignificant compared to how we interpreted the event(s). If we do have self-limiting stories (and I believe if we're breathing, we still have some), then Daniel Amen, MD, neurologist and author of *Change Your Brain, Change Your Life*, gives two helpful probes to disempower those beliefs:

1. Do I know the story to be 100 percent true?

2. What do I know that challenges the story and broadens my perspective?

Clearly, our thoughts, words, and emotions are the driving factors for our actions. Hence we must adhere to the rule:

Notice what you are noticing, for those thoughts will lead to words, ideas, feelings, actions, and our results!

In *High Performance Habits*, Brendon Burchard teaches that outstanding people in all fields of endeavor use situational triggers to provide themselves with reminders for positive thinking and speaking. For example, high performers often use transition time between activities to do a brief mental attitude check. This pattern interrupt allows them to avoid letting unintended energy from one situation to spill over into the next encounter. They silently repeat an intentional mantra, such as "I bring joy." Brendon primes himself each day for positive expectancy. In the shower, he asks himself the following three questions:

1. "What can I be excited about today?"

2. "Who or what might trip me up or cause stress? How can I respond in a positive way from my highest self?"

3. "Who can I surprise today with a thank you, a gift, or a moment of appreciation?"[33]

To sum up this important chapter, I'll use the words of Mahatma Gandhi:

Your beliefs become your thoughts.

Your thoughts become your words.

Your words become your actions.

Your actions become your habits.

Your habits become your values.

Your values become your destiny.[34]

---

33 Burchard, Brendon. *High Performance Habits.* p. 107.
34 Quoted in Lipton, Bruce. *The Biology of Belief.* p. 114.

"If your life's work can be accomplished in your lifetime, you're not thinking big enough."

— Wes Jackson

ACHIEVING YOUR DREAM

Before we get started on Part III, let's briefly review what we've learned so far.

Part I: Conceiving Your Dream Review

In Part I, you examined your six mental faculties of *imagination, intuition, will, reason, memory, and perception*. Until now, you may have taken them for granted without understanding their power. You learned that these Higher Order Thinking Skills (HOTS) make a huge difference in the results you experience in your life. Moving from an unconscious reactionary use of your HOTS to a conscious directed use, you stepped into your magnificent power as a co-creator, choosing the outcomes in this human experience that your heart longs for. You carefully designed each area of the life you would love living, using your *creative imagination* and your *will* to focus. You tested your dream to be sure it is worthy of you—of your focus, your power, and your energy. The primary HOTS needed for this testing were *reason* and *intuition*. Then you brought this dream into clearer focus by amplifying it with your senses and your emotions. *Perception* and *memory* were the HOTS that served you here.

Part II: Believing Your Dream Review

In Part II, you studied some of the universal laws that govern the process of materialization. Going from design to understanding and believing, you fortified your faith that Source is working with you, that *your vision wants you* for a freer, fuller expression of life.

While anticipating your new life, you now began practicing putting yourself inside your vision, *being* your dream by using *imagination* and *reason*.

From this vantage point, you can expect your dream (or something even better) to begin to manifest in quantum leaps. You have examined the formative power of energized thoughts and words—which form ideas and become beliefs that control your actions and results in life. You know to "stand guard" in your conscious mind to choose what thoughts and words you allow to slip into your subconscious "recording" mind. Only thought-words aligned with your vision now receive your nurturing. You made a firm agreement with yourself to let go of limiting beliefs, negative thinking, and low energy words.

Part III: Achieving Your Dream

In Part III, we will move into the very exciting phase of Achieving Your Dream, which is *action*! You'll continue to use your invisible power tools (your Higher Order Thinking Skills) *consciously* to expedite progress. You will recognize that you will face challenges, for this is part of the hero's journey, but you will learn strategies to change these obstacles into stepping stones. In later chapters, we will investigate proven tools to release the remaining ingrained negative patterns that no longer serve you.

"Striving for perfection is the greatest stopper there is. It's your excuse to yourself for not doing anything. Instead, strive for excellence, doing your best."

— Laurence Olivier

CHAPTER 6

MOVING TO MAKE IT HAPPEN

HOTS ALERT: Intuition, Reason, Will

"Nothing happens until something moves."

— Albert Einstein

Remember, we're going from what *is* to what you would *love*—as *transformational* change, not as incremental steps toward a smaller goal. This is exciting because you are actually traversing the gap between your current life and the life of your dream. Through your actions—your behavior—you can change your personality, your relationships, and your future! In the words of W. H. Murray, "Whatever you think you can do or believe you can do, begin it. Action has magic, grace, and power in it!"

As usual, you'll need your five senses to navigate the physical terrain. But to experience an explosive jump, one that accomplishes the necessary physical movement to bring your dream into reality in the material dimension, you will need your Higher Order Thinking Skills. When you harness these amazing mental attributes to perform in your favor, events that seem coincidental will begin happening to propel you forward. Regular practice is required to employ these skills consistently to their fullest advantage, but you are up to the task! Speaking of *this type of thinking* (HOTS), remember Henry Ford said thinking is the hardest work. Earl Nightingale said, "[I]f most people said what they were thinking, they'd be speechless." I mention these assessments by these great leaders not to discourage you, but to accent the point that *you are among the elite 5 percent* of the population who actually study, learn, review, and relearn the principles that require real thinking to assimilate, evaluate, digest, and apply the principles for success. Recognize and applaud your own *will* to grow, to become more, and to become the person you are meant to be. Mary Morrissey echoes Einstein's sentiment at the top of this chapter with her slogan, "Inspiration without action is simply entertainment." So let's get moving!

First, call on your *reasoning* HOTS. Personal development leader Bob Proctor calls this mental faculty our greatest gift. He states that most people operate with a fixed mindset instead of using their reasoning power.

Price Pritchett tells the true story of a life and death struggle that took place at the Millcroft Inn about an hour from Toronto:

> There's a small fly burning out the last of its short life's energies in a futile attempt to fly through the glass of the windowpane. The whining wings tell the poignant story of the fly's strategy—try *harder*.

> But it's not working.

> The frenzied effort offers no hope for survival. Ironically the struggle is part of the trap. It is impossible for the fly to try hard enough to succeed at breaking through the glass. Nevertheless, this little insect has staked its life on reaching its goal through raw effort and determination.[35]

Pritchett goes on to say that freedom is available at the open door—only ten steps away. That possibility would require only a fraction of the energy the creature is expending now. While the fly's current strategy obviously makes sense to it, it is one that will kill the fly, for eventually the small insect will expire on the windowsill.

35 Pritchett, Price. *You²*. Opening page.

In the same manner, we can trap ourselves in "trying harder," while using the same method, the same approach. Einstein said the solution to a problem requires a different mental state from the one that created it. You and I must not rely on taking actions that have proven non-effective, even if we avow to try harder, for we are probably killing our chances for success.

"All right," you say. "I'm willing to *reason* outside the box, to use my *uncommon reason*, to open my fixed mindset to new possibilities, but how do I know what steps to take?"

Your HOTS *reason* and *intuition* will serve you well. Try this exercise:

1. Establish a higher mental vibration by visualizing the situation already worked out.

    Know that there is no problem so great that the Source that created you does not know how to solve it.

    Recognize that you *always* have more to work with right now, where you are, than you are currently aware of.

2. Then take a clean sheet of paper and list things you might do to move yourself forward. (If you have one or two partners who support you in your endeavor, let them participate in this rapid-fire brainstorming.)

    Don't edit; just write down anything that pops into your mind. Stay connected and plugged in to Universal Intelligence.

3. After three to four minutes, review the list. See what holds energy for you.

    Select two or three of the steps.

4. Decide on a completion date and calendar them.

5. As you arrive at next step conclusions, act on them as quickly as possible. The universe loves efficient movement. Observe Nature in this regard. Time, space, and energy are not wasted; metamor-

phosis occurs with expediency. Follow Nature's lead as a template for transformation!

During this initial "what to do next" period, you can begin trusting those hunches, soft nudges, and peculiar ideas (e.g., turn left by that yellow house) that come to you. We'll later discuss in more detail our listening to "the still, small voice," but for now, simply raise your awareness. Notice what you're noticing and what you're feeling. Listen to your thoughts and your body. This type of awareness is the highest form of intelligence, and metacognition will set you apart in your transformation of your thought patterns and vibrational frequency.

The voice of your intuition will be calm and steady, not rushed and frantic. It will resonate with a spot inside you, possibly in your gut. Trust your connection to Infinite Intelligence. Don't concern yourself if you have only a few hunches at first—if the flow is slow. Communication will become much clearer as you build this HOTS muscle *intuition*, and you will become a much stronger receiver.

To reiterate this major strategy of **Sourcing**, let's use the five Ss:

1. **State:** Generate an elevated mental state, releasing any reactive states.

2. **Source:** Pose clear, high-level quality questions, knowing there are an infinite number of possibilities in the universe. Einstein said if he had only one hour to solve a problem, he would spend the first fifty-five minutes forming the question, for the correct question would provide the solution.

3. **Select:** Order the list by priority, choosing the most "alive" and energetic actions.

4. **Schedule:** Put on the calendar the exact time and date your action will take place, and by when you will have completion.

5. **Serve:** Act promptly on the information you have received. Do not be discouraged by current appearances if they don't concur

with your vision. Simply know that this is what it looks like while it is all coming together.

Stop now and Source an answer for a current stumbling block. Use the five Ss on a relatively simple problem to give you practice. Then move to a more complex situation that is blocking you. Remember—you *always* have more at hand to move forward than you think you do. Henry Ford said, "If you think you've done all you can do, you haven't." Mark Twain is credited with putting it this way: "The secret of getting ahead is getting started. The secret to getting started is breaking your complex overwhelming tasks into small manageable tasks and then starting on the first one."

A Magic Morning Routine

In order to take the most inspired actions during the day, you need a morning routine to get you in a peak state. This morning routine is a great way to implement Sourcing Step 1: State. By establishing a routine that puts you in your highest vibration, you set the pace for your movements the rest of the day. You "plug in" to Infinite Intelligence, allowing your *intuition* and other HOTS to operate at peak state.

High achievers understand that this early morning discipline separates the wheat from the chaff. Over the years, I have reviewed the habits of people I admire, such as Wayne Dyer, Deepak Chopra, Bruce Lipton, Mary Morrissey, Eben Pagan, and many others (including my brother, Steve), and each of them has set aside the early rising hours as their sacred time. They use this time for establishing their prime spiritual, mental, and physical states before they begin their day. In other words, they take highly proactive steps instead of falling prey to reactive responses that could derail their intentions. I have congealed the most common strategies used by the top 1 percent of leaders in their field. Of these techniques, I'll give you my personal favorites. We must look first at a few pitfalls to avoid, and then we will scrutinize the best actions to take.

Actions to Avoid

First and foremost, avoid engaging with any social or business media. Many people reach for their cell phone upon awakening, immediately subjecting themselves to a reactive response to the information they receive. In fact, this sort of reliance on input from an external device, such as a phone, iPad, or computer, can become addictive and sabotage your brain's thought process. Research has shown that people lose 30 percent of their productivity by engaging in social media first instead of priming for focused creative activity.

Second, remove yourself from your sleeping environment to a quiet spot set aside for your early morning use. If, like me, you have an occasional battle with your bed and pillow, as to whether or not you should stay, the best way to win your dominance over these wooers is to get up quickly and go to another location. One of my coaches, Benjamin Hardy, has three children, so he actually gets in his car and drives to a solitary point to perform part of his routine! *Whatever you need to do* to set your environment up for the optimum conditions for success is *what you need to do.*

I have a designated room in our home that I call my studio. I restrict the types of activities that take place in this room; for example, the TV is used only as a receiver for mental, physical, or spiritual growth. I don't air current news or cable TV. The pictures in the room suggest creativity and joy to me. The colors are calm and serene. My activities in the room include meditation, writing, painting, and yoga. The music I play has harmonic frequencies that elevate me. While these external "props" may sound minor, the room's energy and vibrations are palpable. It is typical for someone walking into my studio to comment, "I like the feel of this room." I encourage you to give yourself such a spot, even if it is a sequestered corner of a larger room. You will reap the benefits daily.

Three core parts of my morning routine have remained essential to my wellbeing, even though I occasionally modify the list, depending on my different intentions. I have dubbed these three essential parts M^3:

- Meditation
- Mindfulness
- Movement

Let's consider them separately:

Meditation

Paradoxically, to get oneself into an optimum state of movement, the first step is to become still and quiet. Meditation, initially considered an Eastern spiritual practice, has rapidly gained global popularity due to a plethora of evidence-based research showing innumerable physical and mental benefits. Once refuted in Western medicine, it is now considered an alternative way to treat the cause of many diseases. This objective is very different from the pharmacological approach of treating the symptom without addressing the underlying cause of a problem.

Many measurable physical benefits result from meditation. Lower blood pressure, slower heart rate, deeper breathing, improved immune system, and enhanced performance of the digestive system are some noted advantages this practice offers to the regular participant.

Emotional perks are multiple. Lower anxiety, reduced stress, a deeper sense of wellbeing, interruption of negative thinking, stimulation of positive emotions, improved concentration, and overall happiness are all by-products of this practice. These emotions have been correlated with the brain's release of hormones and other chemicals into the bloodstream that flow instantly to receptors throughout the body. For an in-depth review of this process, I highly recommend Candice Pert's *Molecules of Emotion*.

In the early '70s, the Cartesian philosophy prevailed—that only measurable events evidenced by the five senses were considered a valid part of scientific research. The mind's effects were considered irrelevant to the body! The problem that sustained this short-sighted approach was that the world then lacked the technology to mea-sure most of the body and brain's signals, resulting in these millions of signals being overlooked or ignored. As new technology began to demonstrate the fallacies of this separatist view, some physicians began to challenge the ideas they'd been taught in medical school.

As mentioned earlier, in the mid-1980s I served a group of for-ward-thinking internists and gastroenterologists as a biofeedback therapist. We did, indeed, find the above-mentioned physical, mental, and emotional healings for many diseases and chronic conditions. What came as a real surprise, however, was that, de-spite the secular environment of the medical clinics, in addition to the many mental and physical benefits, a spiritual quality evolved spontaneously with regular meditation. After practicing deep re-laxation for a sustained period of 10-20 minutes, some patients "softened" their view of humankind as a whole, while others expe-rienced a connected feeling to all of life. Many patients began fo-cusing on finding meaning in their lives, and others began reach-ing out to people in need. These changes (and in some cases, actual transformations) were unexpected and unusual to us, for there was nothing spiritual in our approach. It seems that when we, as indi-viduals, quiet the many thoughts and voices bombarding us at the rate of 20,000-60,000/day, our calmer sides emerge. I like to think the real essence of who we are then has a chance to peek through.

Researchers throughout the country became eager to learn of our results and to share their discoveries as well. When I had the opportunity to attend conferences featuring some of the leading names in biofeedback therapy and meditative approaches, such as Bernie Segal, Elmer Greene, and Seymour Diamond, they ex-pressed similar observations.

So why are these changes in medical thinking relevant to you now—nearly fifty years later? Because the "treat each body part and each symptom separately" philosophy is still far too prevalent in medical society. The pressures of time, insurance, government aid, expensive equipment, etc., have put many medical practitioners in a compromised position. Additionally, old ideas die hard. Many physicians still cling to the revered concept of a separate body and mind, despite all the evidence to the contrary.

Do not allow someone who holds this fixed mindset to discourage or dissuade you about your own power to work your inner game while addressing the outer game medically to have much better results! Also, do not let uninformed, unenlightened, unobservant individuals throw water on your dream. Using the tools we've examined with the additional set we are about to investigate will bring you success. A critical mass of informed people on the planet are now awake to the infinite possibilities available to each of us. Decide now to be one of that critical mass that causes the "tipping point" for the population as a whole to benefit!

Objections to Meditation

Many people say they don't have time to meditate. Another common complaint is that many find it nearly impossible to sit still. Frequently, people say they cannot clear their mind of thoughts. (To this objection I am compelled to explain that it is a myth that one "stops thinking" during meditation, for that is impossible if you're breathing. Instead, meditation is an opportunity to allow thoughts that slip into your head to pass without charging them with emotion or judging them. This technique is extremely valuable during regular human interactions in busy daily situations, for it builds the "choice factor" for what you will allow yourself to spend your valuable time and energy on.)

To offset the objections listed above of time, fidgetiness, and intruding thoughts, Vishen Lakhiana developed the efficient Six Phase Meditation for the modern world. Instead of attempting to clear the mind, *Six Phase Meditation* guides the brain to structured thoughts that are scientifically proven to elevate one's sense of performance, wellbeing, and spiritual development. They include Compassion, Gratitude, Forgiveness, Envisioning the Future, Your Perfect Day, and the Blessing. Each process is covered in Lakhiani's book by the same name as a system or ritual to use to begin your day. It is available for free on Lakhiani's website MindValley.com. This technique requires only six minutes a day!

My own daily practice is similar, though a bit longer. First, I drink at least two glasses of filtered water to replenish my hydration and to wash away toxins collected in my system during the night. After sitting in my comfortable "prayer chair," I begin meditation. Since I'm a certified Primordial Sound Meditation instructor, I have a bias for mantra-based meditation. I begin by asking the following questions:

1. Who am I?
2. What do I want?
3. How can I serve?
4. What am I thankful for?

Repeating silently a mantra from ancient Sanskrit, I continue this meditation for 15-20 minutes. When a thought comes into my mind, I gently dismiss it without judgment and go back to the mantra. When my chime sounds (I use the popular *Insight* app), I take a few moments to bring my attention back to the room. I then stand and stretch, breathing deeply.

Clients often ask when I meditate. Optimally, I do it twice a day, in the morning when I wake up and in the evening after I recap the day (usually for about ten minutes). The second daily meditation sometimes gets left out, so that is why I advocate the system for morning meditation taught to me by Deepak and Davidji: the RPM Method.

- Rise
- Pee
- Meditate

One of my teachers, Davidji, in his book *Meditation*, gives a comprehensive and interesting study of meditation. I recommend his work for an in-depth discussion of the mental, physical, emotional, and spiritual aspects of meditation. My trainer, Deepak Chopra, has been very influential in bringing meditation to mainstream Western culture. To deepen your understanding of meditation, I heartily suggest you investigate some of his books, articles, and videos available at *Chopra Well*, his outstanding *YouTube* channel.

I have heard it said that praying is talking to God, while meditating is listening to God. There is a strong healing benefit to silence. When I'm still, I can feel my Source, my Creator, as a presence within me, always available to me if only I'll listen. In that quiet, I find a serenity and a calmness that serve me well to reconnect to the intelligent field of consciousness. We are all an inextricable part of this universal field, yet we often ignore it in favor of our "busy-ness." Blaise Pascal has said, "All of humanity's problems stem from man's inability to sit quietly in a room alone."

Mindfulness

> "Drink your tea slowly and reverently,
> As if this activity is the axis
> On which the whole earth revolves.
> Live the moment.
> Only this actual moment is life."
>
> — Tich Nhat Hanh

My mindfulness practice begins with journaling, which has proved formative for me. I write at least three things I am grateful for—selecting things from different areas of my life to reinforce how rich I already am.

I write my vision statement by hand (usually the shorter "travel version").

I then detail my intention for the day; e.g., to be creative in writing another chapter in my book.

I list one or two action steps that will start me toward my goal. Here I also include the emotional charge I intend to bring to the task, such as clarity, levity, inspiration, or motivation. I then mentally activate the primary HOTS I think I will need, such as *imagination, will, or perspective* to create the desired result. In other words, as I journal, I put myself in the position of being the person creating the result I want—priming myself to be fully present when I begin.

To put my subconscious on notice, I briefly note the main challenge facing me so my relaxed mind can begin coming up with solutions.

I also note any "loose" ideas that pop into my head, heeding the advice of Earl Nightingale who said, "Ideas are like fish. If you don't gaff them with a pen, they'll slip away." I don't reread my journals, but occasionally, I glance back at any starred items that catch my attention, knowing that the context for an idea or solution may appear later.

After noting something I love about myself (a habit that took some practice so I could overcome ingrained cultural inhibitions), I read an inspirational passage or Scripture. I spend a few moments in contemplation.

- Is there some weeding of the mind that needs tending?

- Do I need to forgive someone to free space for love?

This stream-of-consciousness journaling only takes five to ten minutes in the morning, and it has proved to be invaluable in helping me crystallize my thoughts, direct my actions, and guard the quality of my dominant mental state. (If I'm the slightest bit grumpy or negative, it shows up immediately in my writing, since I write in a stream of consciousness mode. This awareness of being "off beam" facilitates a quick attitude adjustment, a shift in perspective, which is much easier in this context than it is during an active exchange with others.)

Movement

For the essential third part of my routine, **movement**, I perform 15-30 minutes of physical activity, such as stretching. Yoga is my favorite choice; it adds to my sense of wholeness. Yoga means union. It was developed thousands of years ago to unite mind, body, and spirit. In the United States, we are most familiar with the positions, or *asanas*, but this is actually only one of eight arms of yogic practice. I enjoy the Morning Salutation (which has variations) because it engages all major muscle groups in the body. After three repetitions, I feel centered and invigorated. Yoga is one of the healthiest forms of exercise, and it can be performed at any level of difficulty, depending on the desired outcome and fitness level. Don't buy into the myths about yoga—that it is only for the young and fit, that it requires expensive equipment, or that it's only for women. None of these is true! There are many forms of yoga, including:

- Kundalini Yoga
- Chakras Yoga
- Hot Yoga
- Prenatal Yoga
- Restorative Yoga
- Hybrid Yoga

Consult YogaJournal.com/yoga-101/types-of-yoga/ for complete explanations. A good information source for yoga for healing is www.consumerhealthdigest.com/fitness/restorative-yoga.html.

I also find walking 15-30 minutes a day a good exercise because it requires no equipment, making regularity easier. It is my go-to exercise when no other session is planned.

These basic movements are simply idea-starters. The point here is to include a regular form of physical activity in your morning routine.

The value of consistent physical movement is cleverly illustrated in the following story.

Frogs

Two baby brother frogs were hopping along the meadow lane together when they spied a bucket of cream. With glee, they both hopped into the luxuriously thick cream and began slurping it up in big gulps. When their bellies were filled to the brim, they lazily gazed at each other, laughing about their good fortune. In a few moments, however, those contented smiles turned to wide-eyed panic. Almost simultaneously, the two satiated little guys realized they had a problem—a big problem! The bucket was now totally covered in the slick cream, making the sides too slippery to climb.

"What are we to do?" cried the first brother.

"Hope for a miracle!" said the second.

"Miracles don't happen to frogs," the first said woefully.

Both frogs paddled around in the cream for a bit, but it wasn't long before their tiny legs got tired.

"It's no use," said the first brother as he stopped paddling and slowly sank completely out of sight. A tiny stream of bubbles followed his descent, and then stopped, signaling the end of this little fellow's life.

The second brother didn't know what to do, but he thought to himself that he must keep moving until he could come up with a solution. He paddled and paddled—making circles in the cream, but he still couldn't think of anything. Nonetheless, though his whole body now ached with fatigue, he kept on moving with small strokes. Mama Frog had always told her sons they were more powerful than any situation, challenge, or circumstance. She had told them to keep taking small steps toward their goals, and always to be open to hear or think of new ideas.

Paddling, paddling, paddling. Thinking, thinking, thinking. Listening, listening, listening. Paddling, paddling, paddling.

Finally, as the little frog paused to renew his energy, he was forced to stop, having finally reached his limit. Waiting to sink to the bottom of the bucket like his poor brother, he was shocked to find that he didn't go down at all. In fact, the cream had built up into such a grand mound of butter that he could easily jump out of the bucket to safety! After digging and digging through the butter, looking for his sibling, the younger frog realized his efforts were futile. The frog sadly waved goodbye to his little brother and headed home.

Do you realize that about 95 percent of the population has the very same attitude as the first brother? When the going gets tough, these folks give up. Knowing that movement is essential, you and I *will* keep paddling, using the proven tools we are discovering to become the person of our dream. The obstacles, through continual small steps (or little paddles) will strengthen us. If you're a beginner, start slow and establish consistency.

An object at rest tends to stay at rest. Think of the playground merry-go-round. As all the children hop on, the one designated to start it moving must put her shoulder to the post and push hard. After a couple of laborious circles, the merry-go-round begins to gain momentum, and soon the "pusher" can jump aboard and enjoy the ride. As the device begins to slow down, it only takes a scuff of a toe to pick up the pace again.

In the same manner, a rocket uses the bulk of its fuel to break the pull of gravity—and then it can go for thousands of miles on a little reserve. If we remember that all motion takes a certain shove to get momentum, but that the right kind of effort will eventually become almost self-propelling, we can find the patience to instill habits to build surprising movement toward our goal.

Dr. Paige Tabor posts signs in her office that say:

SITTING IS THE NEW SMOKING.

The research-backed sad news is that our lack of exercise is as big a contributor to chronic diseases as is smoking! Many of our activities promote a sedentary (and deadly) lifestyle. A standing desk is an option that can help overcome the many hours of sitting inherent in our daily activities. A flexible desk (that can adjust up or down) also has many benefits. As Will Rogers said, "Even if you're on the right track, you'll get run over if you just sit there."

These basic movements are but idea-starters for you to select from the plethora of exercise programs available in gyms, as digital programs, and in networking groups. I look for simplicity because consistency is my goal, but your goal might be to up your game in overall physical fitness; then you will, of course, give focus and great effort in this area. Any beneficial exercise will also enhance your mental, emotional, and spiritual sides, so I heartily encourage you!

Summer Smith, Health and Fitness Coach, gives these ten basic tips to keep us moving:

1. Drink water. Consume half your body weight in ounces of water each day to stay hydrated. Add lemon or lime juice to your water before you eat—but *not* after a meal. The acid can help digest the protein you consume.

2. Exercise at least twenty minutes daily. Exercise helps the circulation of our lymph system responsible for expelling toxins from our bodies. Blood flow to the brain enhances a positive mood.

3. Sleep eight hours a night.

4. Lights out! Turn all electronics off at least one hour prior to bed-time.

5. Sleep in a pitch black room. This aids in regulating circadian rhythms. You may also wear eye covers.

6. Kick your sweet tooth habit. Sugar compromises the immune system when consumed. It eats up elasticity in your skin, contributes to tooth decay, and interferes with weight loss.

7. It's important to get natural sunlight so your body can produce vitamin D naturally.

8. Food is medicine. Eat accordingly.

9. Eat a plant-based diet. Use protein as a side dish instead of the main event.

10. Get tested for food allergies. Consuming foods that are allergens for you increases inflammation in your body.[36]

Other morning rituals used by many high achievers that you can try include:

- Take a cold shower. (I sheepishly modify this by turning the water to the cold setting *after* I've finished showering in warm water.)

- Make your bed. This small task helps to organize your mind.

- Spend at least an hour on a large creative project that you're working on; research shows that focus is optimum during the first three hours of wake time.

The point above brings up advice I have used, and routinely share with coaching clients, from several productivity experts, e.g., Eben Pagan, Darren Hardy, Brendon Burchard, Benjamin Hardy, and Deepak Chopra.

36 Source: Summer Smith. Smith is a health coach and personal trainer who submitted this list following a personal interview with me on April 25, 2018.

- Block out your first three to four hours in the morning for putting yourself in a peak state *and* for working on prime projects. Don't check email for 60-90 minutes after getting up, and don't schedule meetings until you have accomplished the work for your own benefit.

Many of the suggestions might require a shift in when you rise. As mentioned earlier, the early bird habit has been the secret to success for many greats, including Benjamin Franklin, Elon Musk, and Thomas Edison (who bragged that the dawn never caught him in bed).

Find your own individual time for peak productivity, and schedule your most important work during this time. Adjust to your natural circadian rhythm to give coherence to all your bodily systems. Famous football coach Lou Holtz said, "In this world you're either growing or dying—so get in motion!"

An Evening Ritual

To magnify the benefits of your "magic morning routine," add an evening ritual. Serving as a "bookend" to bring closure to the day, it also will direct and propel you through your movements the following day. You'll find it an easy method to follow up on unfinished tasks that need to be carried forward. This ritual includes:

1. Shutting off electronic input at least thirty minutes (but preferably an hour) before bedtime.

2. Recapping your day as you jot down points in your journal.

   - What went well?

   - What can you do better?

   By reviewing the events that are fresh in your mind, you will see with much more clarity the solutions and revisions you need to make for the next day.

3. Making your bedroom a secure spot, away from noise and dis-

tress, to allow for optimum sleep. Our society has overlooked the importance of sleep for our mental, physical, emotional, and spiritual health. More on this topic in Chapter 8, but for now, regard your sleep time as sacred to your wellbeing. *Sleep is essential.*

By bringing closure to your day, you set the optimum momentum to move forward and build momentum.

One of the tools my clients have introduced to me is the Panda Planner, a convenient prompt to habituate these different components each morning and evening.

My daily Post-it to you would be as follows:

1. Keep moving during the day—perhaps by setting a timer on your phone to remind you of "moving breaks."

2. Take transition breaks. This idea comes from Ariana Huffington. Before shifting from one task to the next, or one meeting to the next, pause, take a deep breath, and allow yourself to become aware of any tension. Release the tension, and make a conscious choice about how you are going to show up for the next activity. This brief interruption prevents your carrying tension and negative vibrations from one experience to the next. Set your dominant mental state by choice—not by default from residual energy remaining from your last encounter with your environment.

We are all energy—moving in a field of energy. To keep our bodies functioning well as the marvelous, highly intelligent, healing machines they are, we must actually move—in different ways, rhythms, speeds, and durations. To put it simply, being in the flow requires movement!

So, start moving!

Psychologist and productivity coach Benjamin Hardy created the following researched items to maximize performance, which he generously shares with us:

Peak State Checklist[37]

- Have an alarm set across your room so you have to get out of bed to quiet the alarm.

- Spend 2-10 minutes in prayer/meditation.

- Focus on gratitude, and where you are most needed today. This will set your day's trajectory on the things that really matter. It will also provide you with an abundance mindset where you expect good things to happen for you.

- Decide to bring joy with you throughout your day.

- Pull out your journal and write by hand anything that comes to your mind related to your #1 goal or #1 problem you're trying to solve. This will arouse your subconscious breakthroughs you had while you were asleep. As Napoleon Hill has said, *"Your subconscious mind works continuously, while you are awake, and while you sleep."*

- During your journal session, write your big picture vision/goals down in bullet point form and in present tense (e.g. I'm a best-selling author. I'm making over $250,000/year. I'm fully connected and present with my wife and kids, etc.)

- Spend 15-45 minutes in intensive physical fitness [a variation from my gentler approach, but one used by many high achievers].

- Consume 30 grams of protein (plant-based protein powder in water is great).

- Spend 15-60 minutes in focused activity on a big picture goal or passion project (one of those things you've been procrastinating but want to do).

- Take a cold shower. If this shower is immediately following physical fitness, just start with cold. If it is not, start warm and

---

37 Hardy, Benjamin. https://medium.com/@benjaminhardy. Used with permission from the author.

wash your body, then completely switch the temperature to cold for the last 30-60 seconds. During shower, breathe heavily in and out through your mouth. This breathing technique comes from "Iceman" Wim Hof who holds seven world records for withstanding extreme cold. He has a simple method to withstand cold: breathe deeply yet relaxed for 30 seconds, until you can feel what Hof calls the inner fire—a sense of euphoria that extends from cranium to toes.

- Don't check email or social media for at least 60 minutes after waking up.

- After you've spent a few minutes purposefully preparing yourself for HOW YOU INTEND TO BE that day, lovingly interact with your loved ones.

  If you live alone, send a few kind texts to important people in your life.

  If you have kids, play with them for as much time as you can. Be sure to take a few minutes before interacting with them to orient yourself. Don't reactively wake up and then go be with them. You won't be on you're A-Game.

- If you have creative endeavors, spend 60-90 minutes in focused activity on a big project. Research has found that your willpower is highest when you first wake up and your brain is most attuned to creativity. Thus, if you make the time, the morning may be the time when you produce your best and most important work, as is the case for me!

- If you're doing creative "Deep Work" immediately upon waking up, try listening to instrumental/ambient songs on repeat. In her book, *On Repeat: How Music Plays the Mind*, psychologist Elizabeth Hellmuth Margulis explains why listening to music on repeat improves focus. When you're listening to a song on repeat, you tend to dissolve into the song, which blocks out mind wandering (let your mind wander while you're away from work!)

Hardy offers his extensive list as a buffet to choose things from that will work for you. Your list will probably vary according to the season or regarding your current stage in your work. (For example, mine varies when I'm writing compared to when I'm shooting a video, and travel often impacts the order of things.) Regardless of the situation, make the time to plan and execute an effective catapult for your day.

To put yourself in a peak state to begin your spectacular, never-before-lived day, choose from the above morning routine suggestions, or find a similar technique that resonates with you, and begin it *now*.

Chapter 6 is about taking *action*. In order to be sure your actions align with your dream, you must prime yourself each day as you set out to "move and accomplish." At each step of the process, it's important to remember you can take small steps and make little adjustments. Confucius says, "When it is obvious that the goals cannot be reached, don't adjust the goals; adjust the action steps."

Energy can never be lost, but it must be renewed as it is spent in the movements of mind, body, and emotion. Therefore, when scheduling your activities, remember you cannot change the amount of *time* you have, but indeed, you can control how you use that precious commodity. Stephen Covey says, "The key is not to prioritize what's on your schedule—but to schedule your priorities." As you recognize that what you actually manage is your *activities*, you can learn to eliminate those motions that no longer serve you.

Efficiency experts teach that there is more to prioritizing activities than simply listing them in order of importance. The best illustration I know of a better practice comes from famous billionaire investor Warren Buffett. Buffett advocates four steps to productivity.[38]

1. List the twenty-five things you want to accomplish.

2. Prioritize the activities.

3. Circle the top five.

---

38 https://medium.com/personal-growth/warren-buffets-5-25-rule-will-help-you-focus-on-the-things-that-matter-2c383e09d13c

So far—nothing new, right? But—in step four lies the difference.

4. Put the remaining twenty on an "*Avoid at All Cost*" list.

Buffett doesn't expect a person to forego all but five priorities. The key here is to avoid all the other stuff *until you've completed the main five*. In fact, I have heard it said by efficiency experts that if you have more than three priorities, you have none! There are so many things competing for our attention that, in order to succeed at our long-term goals, we must ignore all distractors until we've had success with the primary five. In other words, elimination of tasks is part of movement! We can toss out the old idea that doing a lot of things—being "busy"—is necessarily productive.

Gary Keller, in his bestselling business and leadership book *The One Thing*, gives a clear description of the oft referred to Pareto Principle. This concept states that 80 percent of our profits come from 20 percent of our labor. Keller explains how Joseph M. Juran, a Western Electric consultant, expanded a mathematical model by a little-known, nineteenth-century Italian economist, Vilfredo Pareto. This model for distribution stated wealth was not distributed evenly. In fact, 80 percent of the land was owned by 20 percent of the people. Having noticed that a majority of defects in manufacturing were created by a handful of flaws, Juran suspected this principle might have universal application. In his *Quality Control Handbook*, Juran gave many illustrations of what he humbly dubbed the Pareto Principle. Keller states that this law is as predictable and provable. In his book, *The 80/20 Principle*, Richard Koch defined it well: "The 80/20 Principle asserts that a minority of causes, inputs, or effort usually lead to a majority of the results, outputs, or rewards."[39] Since the minority of our effort leads to the majority of our result, we can readily see how selective focus is key to productivity! He quotes a favorite poet of mine, Shel Silverstein: "But those Woulda-Coulda-Shouldas all ran away and hid from one little Did."

We can then implement new habits—new systems of behavior that

---

39 Koch, Richard. *The 80/20 Principle*. p. 42.

laser us toward our dream. This shift will allow us to bend time in our favor.

In Chapter 8, we will explore more strategies on how to do just that.

First, we must prepare and arm ourselves for the inevitable bumps or even blocks in the road to success. Remembering that no obstacle is greater than the power we have in and around us, we will approach these "wannabe" deterrents as stepping stones to our dream. In Chapter 7, we'll examine and adopt measures to ease our journey, knowing that all we need to see is where to take the next step or two. Scripture speaks of "a light unto my feet." This phrase refers to the metaphor that provides a good visual image of this type of travel. In Jerusalem during ancient times, people traveled at night by foot to avoid the extreme desert heat. The travelers wore oil foot lamps strapped to their ankles to light their way over the rocky terrain. They could not see very far ahead, but they knew to avoid the immediate perils (rocks, varmints, holes, etc.) due to the sphere of light shed right in front of them. In the same manner, we often cannot see the entire path to our vision, but we can determine the next step or two to take in the right direction. If we do find ourselves off course, the foot lamps will help us safely self-correct. In Chapter 7, we will arm ourselves with lights that will enable us to keep moving through challenges.

"Success is stumbling from failure to failure
with no loss of enthusiasm."

— Winston Churchill

ANTICIPATING CHALLENGES

HOTS ALERT: Will, Perception, Reason

"You can conquer almost any fear if you will only make up your mind to do so. For remember, fear doesn't exist anywhere but in the mind."

— Dale Carnegie

Every great accomplishment has been fraught with obstacles to its completion. We humans must have contrast to have meaning (hot/cold, good/bad, alive/dead, empty/full). Without opposites, we just don't "get" concepts, so it is necessary to prepare for the inevitable stumbling blocks that appear to be failures on our way to success. The good news is we are greater than any of our circumstances. We are the only living creatures on the planet that have, in addition to the five senses, a complete set of HOTS. Some other animals have out-paced us in their evolutionary development of their senses to adapt to their environment (dogs smell better with their noses than we do, eagles see better with their eyes, cats hear better with their ears, monkeys have more sensitive taste buds than we, and butterflies can feel gentler breezes). But we are the only lucky creatures gifted with a full set of invisible mental faculties that allow us to go beyond the information that comes directly from our senses to access Universal Intelligence. These HOTS, when properly used, move us forward.

The rough part to this good news is that to move forward, like all great souls before us on the hero's journey, we must slay our dragons of the mind, or at least learn to dance with them. That is no small task, for as the Stoic Publius Syrus said centuries ago, "The pain of the mind is worse than the pain of the body." The primary monster among the obstacles impeding our progress toward transformational change is fear.

Fear

As author Kara Chine states in the blog *Team Better*, "Nothing makes us more aware of our own vulnerabilities than doing something that really scares us."[40] Researcher Brené Brown defines vulnerability as "uncertainty, risk, and emotional exposure." However, this same vulnerability is "the birthplace of innovation, creativity, and change."[41] Chine points out that fear creates problems even beyond the things that scare us. Bowing to fear increases overall anxiety and elevates our reactions over time. "Accumulations of these little mental failures are not good for our self-esteem and confidence. There is only one way out of this conundrum: exposure, habituation, and practice," as detailed by N. Schpancer.[42]

In *The Biology of Belief*, Bruce Lipton tells of his lab experiments that showed, even at a single-cell level, that the mind overrides the local signals from the body. The subconscious has "political clout" in that "she" can send adrenaline with components for protection (contractive) or growth (expansive), despite what your body intends to do. Lipton gives examples of the placebo effect, well documented in scientific literature, as evidence that this phenomenon is also the case in complex, multi-cellular organisms, such as in the trillion-celled humans that we are. The central nervous system— through means of adrenaline—overrides the local signals in the form of histamines from the body.

The subconscious has strong weapons to use against any antago-

---

40 Chine, Kara. https://blog.teambetter.com/something-scary. Sept. 2017.
41 Brown, Brené. TED Talk. "The Power of Vulnerability." June 2010.
42 Schpancer, N. "Overcoming Fear." *Psychology Today*. Sept 20, 2010.

nizer of its homeostatic systems. These derailers are sudden rushes of chemicals that flood the mind and body to stop you short. Problem is, the ever-faithful primitive subconscious midbrain, in her vital role of protecting your survival, has no access to discernment. She cannot understand the nature of the change about to take place, so she does not know whether it's in your best interest. She'll immediately gather up forces and come to your "aid," sending adrenaline and cortisol flowing through your veins, making your stomach lurch, your throat contract, or your palms sweat—resulting in the emotion of *fear*. So what is one to do to prevent this mental and psychological response?

You and I are hard-wired through the subconscious to freeze, fight, or flee if danger is present. (Raphael Cushnir added the reaction "freak out" to the basic three above, and I think it aptly applies to the times we get panicked over nothing.) Without these responses, we would make some awful choices and not last long as a species. Trouble is, as stated above, our older reptilian brain (Lizzie) regards *any* change as scary and dangerous. As she reacts instantly to stop us from proceeding, she'd rather we slide *down* our spiral of growth in order to maintain her version of safety. The midbrain sends an immediate cascade of chemicals to the entire body, leaving us with a yucky feeling in the pit of our stomach or a vague sense of melancholy. Since we can't prevent the subconscious mind's well-intended efforts, let's prepare ourselves to meet the challenges, in whatever form(s) her dragon warriors take.

Opportunity will almost always be accompanied by fear because big opportunities involve risk and, sometimes, danger, as Eben Pagan points out in *Opportunity*. "All mammals (including humans) are programmed by evolution to feel fear when confronted with certain stimuli. We respond in predictable ways to keep ourselves safe, to live another day and reproduce. What's important to know about fear is that it completely changes how you perceive the world. It also changes how you perceive your *own* thoughts and ideas, and how you perceive yourself."[43] Pagan explains that we au-

---

43 Pagan, Eben. *Opportunity* p. 52.

tomatically slip into an avoidance pattern—to avoid loss or death, without engaging our rational minds. "In fact, it short-circuits our rationality, and shuts it down. When we're scared, we tend to act first and think later."[44] If we do think, we instinctively tend to use old patterns that have worked in the past.

The real danger here is that these old patterns avoid novelty and increase our aversion to risk. Today's opportunities are increasingly novel and require innovative approaches, and fear blinds us to these transformative opportunities. Nobel Prize winning psychologist Daniel Kahneman and partner Amos Tversky performed a number of experiments that revealed we humans have a cognitive bias toward *avoidance of loss* over *risk of gain*. This insight, called "Prospect Theory," indicates that people are twice as motivated by loss as they are by gain. This theory is so relevant to behavior change that it is being studied further in a new field called behavioral economics. This observation is important to us Dream Builders to keep us aware of our natural tendencies that might thwart our progress. Jack Campbell puts it best: "Everything we want is on the other side of fear."

Eben Pagan gives a list of the key types of fear we must be alert to:[45]

- Fear of loss
- Fear of conflict
- Fear of rejection
- Fear of the unknown
- Fear of ambiguity
- Fear of missing out (FOMO)
- Fear of death, which he calls the mother of all fears.

Daniel Coleman, in *Emotional Intelligence*, explains the process of flooding, where the mind and body are flooded with chemicals that activate our primitive brain and prepare us to fight, flee, or freeze. It is critically important for us to be cognizant of our own state when fear strikes, to know that our rational thinking has been

44 Ibid. p. 53.
45 Pagan, Eben. *Opportunity*. p. 54.

temporarily hijacked. Pagan reiterates that "making the wrong decision, saying the wrong thing, making the wrong commitment, can be so expensive.… You are basically thinking the way an animal would be thinking if it was faced with being eaten. Making decisions based on the survival instinct will not get you to your goals, as you become strategically blind."[46]

The following story illustrates that through our will we have a choice to override our instinctive fear reactions.

Two Wolves

One evening, an elder Cherokee told his grandson about a battle that goes on inside all people. He said, "My son, the battle is between two wolves inside us. One is Fear. It carries anxiety, concern, hesitancy, uncertainty, indecision, and inaction. The other is Faith. It brings calm, confidence, conviction, enthusiasm, excitement, decisiveness, and action." The grandson thought about it for a moment and then solemnly asked his grandfather: "Which wolf wins?" Putting his hand on the boy's head, the old Cherokee replied, "*The one you feed.*"

We all have these two wolves within us, a dark and a light side (the diabolical and the divine). As we develop the courage to honestly face both of these aspects of ourselves, we gain the power to choose which side to feed. The action steps below are designed to help you pause in the face of fear to feed your faith, your strong belief that you *can* make the necessary choices and that you have the support of a loving Universe behind you.

What are we to do when we find ourselves in a fearful state? As Veronica Roth says, "Becoming fearless isn't the point. That's impossible. It's learning how to control your fear and how to be free from it." Initially, we face fear in all its disguises with coping strategies, and, after becoming calm, we use our cognitive understanding (HOTS).

As a coping strategy, Pagan suggests an NLP technique, a conscious dissociation exercise, to step outside your emotions for a moment.

46 Ibid. p. 55.

Imagine that you're standing behind yourself, looking over your own shoulder at the situation. Hold this mental image long enough to catch a few breaths. As Confucius said, "The man who moves mountains begins with carrying away small stones."

Arianna Huffington uses a similar breathing technique many times during her busy days to dispel fear and stress and to become calm again. She focuses on the rising and falling of her breath for ten seconds, becoming present and centered when she feels distracted, tense, or hurried. Vishen Lakhiani quotes her thoughts in *Code of the Extraordinary Mind*:

> This allows you to become fully present in your life. You know about the thread that Ariadne gave to Theseus so that he could find his way out of the labyrinth after he killed the Minotaur? The thread for me is my breath. Returning to it during the day, hundreds of times when I get stressed, when I get worried, when judgments come up, has been an incredible gift—and it's available to all of us. There's nobody alive who is not breathing.[47]

My own stress relief strategy is similar. Take a deep breath—inhale to a count of four, hold it momentarily, and then exhale slowly (about twice as long as your inhalation), consciously releasing tension and angst in each part of your body. Begin with your forehead, jaw, neck, and shoulders, letting the muscles relax, feeling the contrast. "Rinse and repeat" throughout your body, until you feel a calmness, followed by a surge of power. This breathing technique is effective for Dragon 1: *Fear* because it brings you back to the present, putting a halt to your imagination's panic, as Eckhart Tolle teaches in his life-changing book *The Power of Now*.

Here's author Sonia Choquette's advice for when sudden anxiety creeps in: Be present. Take your attention away from the fears circling your mind and notice the sensory input at hand in your environment. Find the beauty in the intricacy of the markings on a leaf, the textures around you, the soft fragrances in the air, and the

---

47 Lakhiani, Vishen. *Code of the Extraordinary Mind*. p. 184.

different sounds you hear. This level of consciousness forces you to take your emotion off the fears, worries, and anxieties so you can remember who you are and concentrate on the now.

A related aid to use in the heat of battle is one of our HOTS: *will.* Decide to focus, change the direction of your thoughts, and be still. Counter-intuitive? Indeed it is, but here we dip into the infinite dimension of our Universal Intelligence. "Be still, and know I am God." This line of Scripture speaks to the calm, loving, peaceful place of respite available to any harried person.

It's that feeling of awe and amazement one gets when gazing at a full sky of stars. Yes, I am a tiny speck in the universe, but I am also an infinite being, supported by—and an integral part of—all of this creation. "I will fear no evil, for Thou art with me" (the Twenty-Third Psalm). Famed astronomer Neil de Grasse says that star-gazing makes him feel immense, an inextricable part of the universe, for he knows we are, in fact, made of the same stardust, the same elements of carbon, nitrogen, and oxygen, as the infinite number of galaxies above.

Once you have used your will to focus on stillness, the antidote to returning to the frightened state is to come *from* your dream. Put yourself inside the frame; see yourself *being* the person you are aspiring to be. This action requires strong feeling. The strategy is not one of escape; it's a shift in *perspective.*

During World War I, Teddy Roosevelt often stood on his veranda and stared at the stars. "That's a spiral galaxy in Andromeda, which is as large as our Milky Way. It's one of a hundred million galaxies." He'd repeat this fact over and over until he finally could say, "Now I've made my problems small enough." What a beautiful example of using *will* to stop and focus and to gain a shift in *perspective!* With practice, we can often recognize our problems aren't as grave as first perceived.

Napoleon Hill insists in his seminal book *Think and Grow Rich* that "Thought forms of fear of poverty cannot be transformed into thoughts of courage and abundance."[48] I mention this type of fear for it crops up as a shadow of the mind when it's time to make a large commitment to our dream. Because the only thing we have control over is our thoughts, Hill tells us the choice is clear: We can let the thoughts of others influence us, *or* we can guard our thoughts closely and entertain only those thoughts we choose to energize.

By killing two of the most essential HOTS to accomplish new ideas, *creative imagination* and *uncommon reason*, fear of poverty can completely derail action. The Great Depression is a good example of what happens when the fear of poverty is allowed to solidify.

The roads of abundance and poverty lead in two opposite directions. By overcoming fear, you are creating a mental attitude of riches. Remember, everything is created twice; once in the mind and then in material reality. Wallace Wattles states that when we realize there is but one Thinking Substance *from which* and *by which* all things are made, we quickly lessen our fear and doubt.[49]

To dispel the vicious and sneaky fear of poverty, Napoleon Hill gives this proven method in six steps:[50]

1. Fix in your mind the precise dream you desire. State the exact amount of money you intend to manifest as part of the abundant life of your dream.

2. Determine exactly what value you intend to give in return.

3. Establish a definite date when you intend to acquire those funds, and—even more importantly, become the person in your dream.

4. Create a definite plan—or a specific series of next steps—to put into action.

5. Write a clear, concise version of your dream with the time limit for its manifestation.

---

48 Hill, Napoleon. *Think and Grow Rich.* p. 26.
49 Wattles, Wallace. *The Science of Getting Rich.* p. 17.
50 Hill, Napoleon. *Think and Grow Rich.* p. 22-23.

6. Read your written statement with great feeling each morning when you awake and each evening just before going to sleep.

To clear my consciousness of congested fears, I find it beneficial to write my vision by hand each morning to ground myself in my new reality and to plant fertile seeds of faith. I remind myself that a rich life includes abundance in all domains of life's expression; feeling this wealth that is already mine frees me from the stress of attempting to achieve.

Outright panic is easy to spot in ourselves or others. But fear often comes in subtler forms, in disguised modes. Let's look at two of the most common by-products of the fear monster: doubt and delay.

Doubt

Dragon 2—*Doubt*—involves those self-deprecating obsessions. Am I good enough? Up to this? Tall enough? Funny enough? Smart enough? Pretty enough? You get the picture.

Most creators have met with doubt from those whose minds cannot conceive the invisible, who cannot dream beyond "ordinary reason." The Italian inventor Guglielmo Marconi faced this demon in his outer world. Ignoring Marconi's inner convictions, his family and friends so doubted his aspirations to invent long-distance wireless radio communication that they sent him to an insane asylum! Despite their doubts, years later people around the world listened to their radios in rapt attention as Rear Admiral Richard E. Byrd communicated live from Little America, an Antarctic exploration base and his and his team's campsite, as their expedition approached the South Pole.

We can all be grateful that thinking minds ignore the doubters and focus on their inner vision. This focus requires determination and persistence, as well as fortitude and faith. There are no shortcuts.

In my early thirties, I married a man who had a gift for making others feel good about themselves. I thrived during our years of marriage, often feeling I was standing on his shoulders, seldom doubting my value (a big departure from my past). After his death, I was consumed with self-doubt. I discovered the hard truth: When we attach our sense of self-worth and happiness to someone or something outside ourselves, we become dependent on that outside source for strength.

I had to learn anew to love myself and others from my own inner self. Chopra teaches, "Once you know who you really are, being is enough. You feel neither superior to anyone nor inferior to anyone and you have no need for approval because you've awakened to your own in-finite worth."[51] This advice proved invaluable, for internal confidence is the only reliable source of strength. No longer doubting my pow-er to grow, and benefiting from the unconditional love I had experi-enced, I found I could generate love for myself as a unique spiritual being in this precious human experience. It became easier to set goals for myself without modifying them because of disapproval from oth-ers. Having experienced unconditional love for my two daughters, I learned to expand that love to others. I set internal standards for my own life, based on my values, independent of those around me at the time. Recognizing the road ahead would be fraught with obstacles, I determined that I had the capacity to deal with them. To paraphrase Eleanor Roosevelt, I learned that all the water in the world could not drown me unless I let it get inside me.

The state of being that best describes what I sought (and continue to seek) is unfuckwithableness. That word and definition showed up on the Internet in Wikipedia in 2015.

Unfuckwithable: A definition

When you're truly at peace and in touch with yourself. Nothing anyone says or does bothers you, and no neg-ativity can touch you.

Have I totally reached this state? No—but I *am* becoming more im-mune to outside criticism *and* praise. By bookmarking each day with

51 Chopra, Deepak. *21-Days of Inspiration*. p. 16.

my own expectancies and wins, as well as my outcomes that need improvement, I more easily recognize that this journey is a process. Despite circumstances, I have an inner source of striving—a longing for growth that excites me, and a deep feeling of happiness that comforts me—even in the worst of times. I give and receive unconditional love. Am I consistently "unfuckwithable"? Do I always avoid letting doubt slow me down? No—but just watch me grow!

I challenge you to do the same. Become totally grounded in who you are. Recognize and use your HOTS to control your mind and feelings. Dale Carnegie teaches us an important lesson: "Inaction breeds doubt and fear. Action breeds confidence and courage. If you want to conquer fear, do not sit home and think about it. Go out and get busy."

Napoleon Hill gives these three keys to erasing doubt:

1. Visualize yourself as successful, capable, and confident, using your HOTS *Imagination*.

Use every positive emotion you can muster to delight in imagining yourself as the person living the life you are designing. Bring in your senses, just as we did to "cellularize" and "sensorize" your vision in Chapter 3. Your subconscious will respond to this input by flooding your brain with endorphins and serotonins which will neutralize the doubt. The doubt muscle in your mind has grown strong over the many years of repetition, so frequent practice of the new image is necessary. What fun to explore your powerful, sexy, vivacious self on a regular basis!

2. Reflect on your past successes. (HOTS *Memory*)

We humans have a negative bias in our memory because we easily recall our failures (aka feedback) and minimize our wins. It is important to bookcase each day with a recapitulation (as my mentor David Simon called it) of our successes, no matter how small. In addition to noting how to improve the day, we must balance our recap by recognizing and celebrating the positive changes we make along the way. If we applaud our baby steps, the giant leaps become more plausible, for we gain confidence from our new behaviors.

3. Have clear direction with specific goals, and self-correct imme-
    diately when you discover you are off track, using your HOTS
    Reason and *Perception*.

A rocket propelling into space to land on the moon is slightly off
course more than 95 percent of the time, but the airship quickly
corrects its trajectory toward its target. In the same manner, we
must take our next step—often before we feel "ready"—and remain tuned in to our inner knowing to shift as needed. As Dr.
Martin Luther King, Jr. said, "The ultimate measure of a man is
not where he stands in moments of comfort and convenience, but
where he stands at times of challenge." When it's hard and you begin to doubt your acumen, press on. This is the point where your
small, consistent steps toward your clearly-defined target will have
built a platform for you to stabilize, gain momentum, and make a
leap that will separate you from the crowd.

Delay

To propel your forward momentum against Dragon 3, *delay,* aka *pro-
crastination*, I offer a simple tool that has worked for me against this
persistent boogeyman. Each night, before retiring, write two or three
tasks that will bring you closer to your vision. To select these remark-
able steps, we'll use Mary Morrissey's acronym: **ELFS**.

•   Easy: Decide if it's easy—that you are moving in the right direc-
    tion because this action aligns with your goals.

•   Lucrative: Choose something that has a *lucrative* potential; it will
    enhance your position for making money.

•   Fun: Make sure it's fun; Spirit loves fun.

•   Serves: Select a task that serves you and others involved.

After your Magic Morning Routine (as discussed in Chapter 6), complete at least two of these ELFS tasks. At times, we must be willing to

do menial tasks if they provide an opportunity to move us closer to our most cherished goal. Capitalize on your momentum to keep going. Add to your list the following night.

When other To Dos come to mind that don't fit your ELFS criteria, put them on a separate list, keeping these four special ELFS as priorities.

By engaging in this process, you are adhering to Napoleon Hill's observation and experience that, in order to achieve your dream, you must have definiteness of purpose, a clear knowledge of what you want, and a burning desire to possess it.[52] Hill states that the reason you become "master of your fate and captain of your soul" as W. C. Henley wrote, is that you learn to control your thoughts. He gives examples of great thinkers who changed the course of the planet by fixing in their minds the objects of their passions and resolutely ignoring others' despairing opinions.

Henry Ford's engineers declared the V-8 engine impossible for months and months as they tried every conceivable way to carry out Ford's orders. Ford persisted, and so did his engineers (they had to in order to stay employed), until finally the solution appeared!

The Wright brothers had many, many failed attempts to get their aircraft to stay in the air for a sustained period. Not only were they criticized, but they were also harshly reprimanded by some churchgoers as heretics, for these people feared that man should not fly. Fortunately, these imaginative brothers prevailed, and changed the course of human travel forever.

In the late twentieth century, Bill Gates had the vision of making computers a valuable tool for every home and office. He dropped out of Harvard University to devote all his energies to the development of a software company with his childhood friend, Paul Allen. They changed the course of information exchange for everyone.

I chuckle to think of my own admonitions to a friend in college. Johnny Deutchendorf was a buddy of mine who made group outings at the lake much more fun because of his great music. A shy

52 Hill, Napoleon. *Think and Grow Rich*. p. 24-25.

but very likeable fellow, Johnny played the guitar and sang with a three-piece band. Having attended the same high school, he and I often chatted during fraternity parties. He dreamed of leaving school and becoming a singer and songwriter. In his sophomore year, he decided to follow his dream. I chided him that his success in music was not guaranteed—that his future success depended more on his completing his college education. Johnny left school that year, and I lost contact with him.

One evening, years later, I recognized a familiar voice on television. Racing into the living room, I saw my former friend hosting a Christmas special on a major network! He looked much the same, but he was now being announced by his stage name: John Denver. The backdrop on the set had spelled out his family name—so my exclamations of wonder at seeing Johnny Deutchendorf were instantly validated. Having had no idea that several of the top hits I loved were his songs, I humbly laughed at my earlier well-intended doubt and argument for delay.

John Denver's quest points to the fact that each of us will encounter doubters as we begin a transformative life change. The time is never "just right" to begin—with all the loose ends tied up neatly. However, when your dream stands the tests we've covered, when your heart pounds with longing for that dream, when you hold your vision with clarity and work with "right thought and right action" (as Wallace Wattles called these principles), success will follow.

Two important concepts that can alleviate the fear dragons are:

1. These feelings and hesitations are good news, for they let me know I'm on my "green growing edge." Ultimately, I'm aiming for quantum leaps, not simply incremental steps, and this surge will always require my stepping out of my comfort zone, hence alerting my subconscious to fire chemicals throughout my system. If I'm not feeling a bit scared, my dream isn't big enough to warrant this huge effort.

2. Roadblocks often become stepping stones, as they require me to

"up my game" in my Higher Order Thinking Skills. I must peak my imagination, listen to my intuition, use my uncommon reason, flex my will to focus, enhance my memory, and—most of all—shift my perception of what is possible, probable, and predictable. Henry Ford said even a mistake may turn out to be the very thing needed to create a worthwhile achievement. He called life a series of experiences, each one of which makes us stronger, even though we may have trouble recognizing that fact at the time. Edison stated that one of the greatest discoveries a man can make is when he discovers he *can do* something he previously feared to do.

Similarly, Ford advised, "When everything seems to be going against you, remember that the airplane takes off against the wind, not with it." Thankfully, the Wright brothers and other great minds understood this principle, or we would still travel only at the speed of the fastest horse. Without these challenging dragons to overcome, without the drag they create that gives us courage to lift our noses toward the sky, we could not soar to our vision's heights.

Be sure to read additional motivational material by or about people who have demonstrated success, such as Herb Kelleher and Rollin King, founders of Southwest Airlines, who bucked the whole airline system to provide affordable fares for air travel by eliminating unnecessary services. In an industry with losses in the billions, Southwest Airlines (which began with an idea on a cocktail napkin) has an unbroken string of thirty-one years of profitability. Yet the industry transformation displayed by this company came not from its profit margins, but from the high-performance relationships the leaders created for an enormous competitive advantage in motivation, teamwork, and coordination. I have used the inspirational book *The Southwest Airlines Way* by Jody Gittell as a training tool for leaders in several fields.

In *You²*, Price Pritchett takes a no-nonsense approach, revealing that a quantum leap—a jump that does not stop at all the bases or even follow the usual confines of time—is no more difficult to achieve than is forward movement in incremental steps. He says it is necessary to destabilize yourself as you cut loose from the familiar and truly test your own limits. Pritchett insists that if you are not experiencing challenges, you have not set your vision high enough. Only by making mistakes can you learn what works, and what you are capable of.

The major obstacle to overcoming stumbling blocks, Pritchett says, is overcoming your own resistance to challenging the risks. Without challenging them, by accepting the status quo, you are robbing yourself of the chance to make a big leap that doesn't follow the traditional constraints of distance and time. You actually *give* yourself a big chance when you decide to *take* a big chance. You are not gambling; you are opening yourself to new possibilities that you have been ignoring. By testing your limits, you actually begin to turn the odds regarding what you can achieve in your favor. Positive action is what makes the difference. By taking the offensive, by leaving your comfort zone, you cause your dream to crystallize into reality. My coach, Dr. Kirsten Wells, teaches that "Fear is the energy of the border of our conscious awareness, the reality that we know." When, as a student, I reported encountering a large obstacle, she would say to me, "And that's the good news. You are standing on the green growing edge of your awareness!"

John F. Kennedy put it well when he said, "There are risks and costs to a program of action. But they are far less than the long-range risks and costs of comfortable inaction." Without moving past the apparent risks, you will never know what you lost when you agreed to accept the status quo. Leonardo da Vinci puts it this way, "Obstacles cannot crush me. Every obstacle leads to stern resolve."

We have looked at several approaches to overcoming challenges. One thing is certain: challenges are to be expected. They *can* be turned into stepping stones with the right mental state and plan of action.

When facing a challenge, I read this poem from the ancient poet Rumi to remind myself that each visitor has a purpose:

The Guest House

This being
human is a guest house.
Every morning a
new arrival.
A joy, a depression,
a meanness, some
momentary awareness
comes as an
unexpected visitor.
Welcome and entertain
them all!
Even if they are a crowd
of sorrows, who
violently sweep your
house empty of its
furniture, still,
treat each guest
honorably.
He may be clearing you
out for some new delight.
The dark thought,
the shame, the
malice. Meet them at the
door laughing and invite
them in.
Be grateful for whatever
comes because each has
been sent as a guide from
beyond.

Decide on an approach you want to use to armor yourself, and make clear note of how you will implement it when you do encounter a problem. Think of different situations that will call for different actions (versus reactions).

Take time right now to jot down some notes to remind yourself of the steps you will take when fear, doubt, or delay strike. If you allow these obstacles to pile up, they will become your regrets of a lifetime. So—big or small—take the next step. Don't fear uncertainty; fear mediocrity! When put in motion, an object normally stirs up dust. Get ready to kick up some dust clouds as you continue to take your action steps! Keep in mind that your fear is a good indicator of the next thing you should be working on. Know that you, the hero, can face the dragons with certainty that you have the power of the universe working in your favor!

_____

_____

_____

_____

_____

"Motivation is what gets you started.
Habit is what keeps you going."

— Jim Rohn

CHAPTER 8

ASSUMING LEADERSHIP OF YOUR MIND

HOTS ALERT: Perception, Reason

"It will be difficult to break the habits of thinking Abnegation instilled in me, like tugging a single thread from a complex work of embroidery. But I will find new habits, new thoughts, new rules. I will become something else."

— Veronica Roth

In Chapter 7, we discussed the predictable obstacles that often crop up in the journey of growing, transitioning, and transforming. In this chapter, we will go deeper into the proven methods not only to overcome our fears, distractions, and delays, but also to propel us in the direction we want to go. I will use two terms coined by Vishen Lakhiani in his outstanding book *Code of the Extraordinary Mind*. They have resonated with me and with my clients so well that they are now part of my coaching vocabulary. After a brief description, we'll raise our awareness to the impact of these terms.

Culturescape. Our *culturescape* is the world we live in, the one that has shaped us. As the name implies, this term includes all the influencers in our environment and in our cultural history: our parents, our teachers, authority figures, the media, religious customs, and social mores of all types. I like this word's encompassing quality because we cannot underestimate the profound influence the many ideas we have absorbed from our environment (that have gelled into conscious and unconscious beliefs) have had upon us. Our actions and the subsequent results in our lives reflect these belief models.

Brules stands for *bullshit rules*. Its clarity of meaning is the appeal for me. Our beliefs dictate the rules we follow—our systems, our how-tos for everything we do. As Steve Jobs pointed out, "Most rules are made by people no smarter than you." Faulty beliefs dictate ineffective rules—hence Brules.

Some of the rules we adhere to give a sense of order or bring enriched meaning to our lives. However, many rules have become outdated; they no longer fit our circumstances or apply to who we have become. These Vishen calls *Brules*. To become authentic in expressing who we are and in living our purpose, it is necessary to decide to let go of those obsolete mandates that no longer serve us.

Attitudes that denigrate others because of general Brules about race, religion, and beliefs must be discarded to allow each of us to expand our awareness of the beauty of life. Faulty, derisive, and outdated beliefs must be replaced by empowering, compassionate, and inclusive ideas that allow for synthesis of purpose for all of us. Concepts around sexuality run rampant around Brules, so accept my recommendation and earnest persuasion to examine those closely to be certain they align with your own core values.

Lakhiani gives a helpful analogy by comparing our beliefs to computer models. When our PC or Mac becomes obsolete, we update the model so we can receive and process information efficiently. Our habits Vishen likens to the software systems we install on our computers. If a program lacks problem-solving ability, has limited efficiency, or "freezes up," we replace it with a better version or a completely new system. In the same way, we must regularly scrutinize both the models and the systems constantly working in both our conscious and subconscious minds. Overused, outdated habits dramatically slow down our processing speed and our progress. Obsolete models can derail us completely in our transformation. Thus it is our responsibility to use our HOTS *reason* to check out the facts in Brules and evaluate the efficiency of our models (beliefs) and systems (habits).

Conscious reworking and rewiring of persistent negative thinking and chronic sloppy performance standards will yield exponential gains in becoming the *you* in your dream. It is critical to employ your HOTS *will* with the HOTS *reason* to sort out those weeds in the fertile soil of your mind; then you can nurture the tender young seeds of your big bodacious dream.

In this chapter, we'll explore proven techniques to overcome the deeply seated models of behavior from our culturescape that no longer serve us. Then for our own evolution, we'll examine some of the Brules—the systems many of us have blindly followed without recently questioning their validity. It's good to remember that some rules were established as perfectly good safeguards at the time of inception, but with changes in circumstances, they no longer serve their positive purpose, and they can delay or even destroy our progress. We can cast a kind backwards glance to this type of Brule, but then eliminate it from practice.

Napoleon Hill says discipline of the mind comes through self-control. He tells us before you can control conditions, you must control yourself. Self-mastery is the hardest job you will ever tackle. "If you do not conquer self, you will be conquered by self." Keeping your focus with meditation and "good thinking" (Wallace Wattles) is key to taking charge of the thoughts, words, emotions, feelings, and ideas that bolt into your mind at lightning speed. Good thinking, just like our gardens, requires close inspection and weeding. Are the "plants" healthy, or is some form of destructive insect feeding on the leaves—or on the roots? Does the soil around the growth support or inhibit it? Do some of the plants need pruning? Has enough light shone on the garden to see what is beneficial, what is overcrowded, what is rotting, and what is even spreading spoilage? Have errant seeds taken up residence? Just as the avid gardener knows the careful evaluation that must be done, the custodian of the mind also has the task of examining beliefs that may or may not be obvious at first glance.

This process then extends to the habits formed from those imbedded beliefs. Are they effective, or do they need to be updated or even deleted? The importance of this question is clearly stated by Tony Robbins, internationally recognized leader in personal development: "Our beliefs are like unquestioned commands, telling us how things are, what's possible and impossible and what we can and cannot do. They shape every action, every thought, and every feeling that we experience. As a result, changing our belief systems is central to making any real and lasting change in our lives."

In the *Code of the Extraordinary Mind*, Vishen Lakhiani gives many examples of beliefs that have become part of our life as we have lived with models from our parents, teachers, friends, and even societal thoughts that have gained strength. For example, during the stock market crash of 1929, the pervasive feeling of loss led to an entire society's belief in The Great Depression, which lasted until new thought patterns of growth and rebuilding could take root. Again using the metaphor of computers to clarify these concepts, beliefs are the models, framework, and hardware of our thinking. When a computer can no longer handle the latest apps or run the newest operating systems, we expand its capacity or get a newer model. Our habits are like software. If our new computers have more speed and capacity than the software we are using, we update the program or delete the app and upload a better version. Similarly, we all must be vigilant to guard against outworn models that create outdated habits, and we must protect our work from the many "bugs" and viruses that can sneak innocuously into our minds.

Some of the models to inspect first are those that will impede our transformation. For example, biases for or against age, gender, race, ethnicity, religions, and physical appearance all crop up unexpectedly through experiences that trigger our emotions. Then—

without our recognition—these feelings coalesce into actual beliefs. How does that happen?

You and I have "meaning-making machines"—our brains. The conscious brain, called the prefrontal cortex or the thinking brain, gathers input and assembles it into a pattern that resembles something already known or experienced. It communicates neurochemicals to our body that translate for us as feelings. Without all the facts, without the altitude to see the whole picture, and with neurological tracks that have become dominant through extensive use (wear and tear), the conscious mind instantly pulls in many pieces and calls the puzzle solved. The ever-vigilant subconscious mind wants to protect us. To do so, she makes a record of events as they happen. If the event is traumatic, she carefully secures the memory of it in long-term memory (the cerebellum), where it can stay long after the crisis has passed, but still holding a firm grip on the actions Lizzie will allow. If the memory is a feeling of inadequacy, helplessness, or weakness, it becomes a self-fulfilling prophecy for the traumatized person's subsequent actions, even if the situation was resolved.

This response can be perpetuated as patterns begin to create anticipated results. For example, we think the world might treat us unfairly because that is what seemed to be the case last week, and we notice—with our ever-vigilant radar—our reticular formation in the brain—each and every instance that fits this pre-ordained criterion. This repeated attention is how, over time, unfair treatment becomes our faulty-thinking's result. As we settle into our societal norms and think "that is just the way things are," we stop having "peak experiences." Dr. Abraham Maslow says these peak experiences are necessary to live fully as a self-actualized person. Most of the population gives up on changing their expectations, instead of striving to create novel situations and new experiences.

What if we could challenge those beliefs to see if indeed they hold any validity? We definitely can! Welcome, *uncommon reason*, a very special human HOTS. I say *uncommon* because it can become a habit to strap ordinary reason on the back of our prejudices (aka untested beliefs) and use this limited, restrictive perspective to make choices, to decide what is true or untrue in our own reality. When we decide (which, you recall, means to cut out) to eliminate those beliefs that don't stand up to calm, reasoned scrutiny, we gain an expansive *perspective* (HOTS) that opens us to new, creative possibilities.

In my own life and teaching experience, I have found that we hold two kinds of beliefs, each of which is critically important:

- **Beliefs about the world we live in:** That's the stuff we "know," like how the economy works, and how we should earn a living, raise our children, behave in the workplace, sustain our family unit, socialize, respond politically, and even have sex. All the other possibilities that impact us outside ourselves fall into this category.

- **Beliefs about "me"—who I am:** How do I feel about myself? How do I think others perceive me? What is important to me? What trait would I change if I had a magic wand to tap a new one into existence? What would I keep the same? How much risk is good for me? What keeps me safe? What kind of power do I have? Am I part of something larger than myself? Do I recognize my own importance, or am I part of life's flotsam? Am I sexy? Am I strong physically? Am I enough?

All of these questions have answers in our inner beliefs that control 99 percent of the actions we take. This is by no means an exhaustive list. I invite you to probe deeply and find the questions and answers that you hold about yourself. Decide whether they sabotage or support your growth. We tend to keep these beliefs about ourselves tucked away out of sight of our conscious thinking, while our beliefs about the world in general are sometimes a bit easier to detect—especially when they are exacerbated by external events.

There are forums to discuss gender issues, ethnicity, politics, economics, climate change, philanthropy, etc. There are fewer opportunities for the safe, sacred space to delve into the hidden beliefs about yourself. I believe that is why personal coaching has grown so rapidly. With so many novel opportunities bombarding us, many of us seek a focused, protected time to explore so we can actually discover our beliefs without assuming there is a pathology or disorder involved. Problem or challenge—yes. Illness—no. Just as athletes have, for many years, hired coaches to improve their performance, most high achievers use life coaches who are trained to support, motivate, inspire, and help them pattern personal development at a faster rate.

The media today is filled with stories of challenges to long-held societal beliefs or customs. Topics of gender and ethnicity equality fill the air and wires as the institutional norms are beginning to crumble. While advances are being made, we, as a "civilization," have a great distance to go. Only through individual openness and introspection can we reach a critical mass to eliminate the ignorance that holds so many of us captive. In addition to public concerns, more subtle areas exist that seldom receive attention; we'll focus on some of those now.

Cooperation vs. Competition

The Darwinian theory of "survival of the fittest" has become so pervasive in commerce that it often goes unnoticed. Seemingly harmless at first, the competitive forces at play do not bring out the finest or the most beneficial in us. I am referring to the "for me to win, you must lose" syndrome.

Evolving now are some businesses whose mission is to grow and excel, but in a manner compatible with other businesses, beneficial to the population as a whole, and sustainable for the earth. As our planet becomes more and more depleted of resources, we must reexamine our concept of economic success.

The old separatist patterns of thought—the "dog eat dog" approach (mentioned above)—according to Bruce Lipton, is not the natural order of living species. In his laboratory, Lipton found that single-celled organisms—having all the major bodily systems that we have (respiration, circulation, digestion, and elimination)—colonize when food supply becomes scarce. They react to a limiting circumstance by joining together to become a more complex organism. Remarkably, the cells then take on specific functions, such as becoming part of an organ to serve the system as a whole, according to their proclivity for that type of work. They become an integral part of the new organism—with far more sophisticated means of adaptation and survival. Here is a brilliant example of how we, as highly complex human beings with trillions of cells, are designed by Nature to join together in cooperative, win-win situations. This shift in mass consciousness from competition to cooperation must become more widespread to protect and evolve our human species, all life forms in general, and our beautiful planet earth.

Forgiveness vs. Revenge

A generally held and widely promoted societal belief for us to examine is that of the premise of "an eye for an eye." This concept becomes confused with justice—rightful action—when it takes the form of vengeance. For centuries, outdated beliefs have promoted many battles in the name of religion. Movies and television dramas glorify the motive of revenge. Despite all these promotions of holding resentment, grudges, and hostility, the only healthy way to overcome having been wronged by another is to forgive.

In a *National Geographic* documentary, Morgan Freeman interviewed Paul Kagame, the current president of Rwanda. Following years of civil war, and a plane crash killing the Hutu president, members of the Hutu majority government participated in a mass slaughter of approximately 800,000 Tutsis and Hutu-moderate civilians. The devastation was overwhelming, but ignored by the global community.

Paul Kagame worked tirelessly to teach the Rwandans that they could live in peace. In his interview with Freeman, he stressed that justice and revenge are not the same, stating:

"Revenge keeps the victim trapped, never allowing the wronged person to move on."

The startling part of the documentary was that Kagame has established communities in which the Tutsi survivors and their Hutu assassins live together, helping each other in friendly cooperative work. As these former enemies communicated over time, they grew to know, understand, and support each other. I have witnessed no stronger image of a symbol of hope for lasting peace. Most of our betrayals are not of this magnitude, but they still come with a hard price.

Forgiveness is not easy. Betrayals hurt deeply. However, to clear our fertile minds of the weeds of bitterness, it is necessary to let go of these potent negative feelings that choke out the seeds of our dreams. Buddha says holding a grudge is like holding a burning charcoal in your fist, planning to throw it at the person you resent, and expecting it to burn the other person. Resentment is like drinking a sip of poison every day, expecting the betrayer to die.

In *Six Phase Meditation* (available for free at MindValley.com), Vishen Lakhiani teaches a technique to grant forgiveness:

1. Pick something easy first.
2. See this person in front of you.
3. Feel the pain they caused you for 30 seconds.
4. Ask yourself why they did what they did.
5. Think about what you could gain from this experience.
6. Forgive into love.

To forgive *yourself*, see the younger version of yourself, and apply the same techniques.

The process for forgiving others that I learned from Bob Proctor, self-development coach, is similar, but it has an important additional step that I find extremely useful when I have a deep wound that is difficult to release:

• Hold the image of a person who is close to you, such as your child, in your mind, letting your loving feeling fully emerge.

- When your body is deeply immersed in feelings of love, bring forward the image of the person you are attempting to forgive. There will be a sudden contraction of feeling as you shift images.

- Go back to your loved one and allow your heart to expand.

- Repeat this process several times until finally the sharp hurt subsides as you bring up your betrayer.

- End with a vision of love and gratitude around your dear one.

Perform this ritual daily over a month or two as required, until the negative impact has dissipated.

Jim Hardt at the Bianaut Institute examined thousands of people's brain waves while in deep meditation, discovering that the most reliable way to obtain heightened brain states was through forgiveness. Vishen Lakhiani reminds us in his *Six Phase Meditation* that forgiveness of another person is not excusing him or her of past behaviors. It is freeing yourself from holding a grudge, thereby making yourself responsible for your past.

To forgive is to "give for" freedom, as taught by Mary Morrissey in *DreamBuilders*. Mary tells of a grocery clerk who was very rude to her. When commenting about her displeasure with the encounter to the grocery boy who carried her bags to the car, she learned that the checker's son was in the ICU. The mother was unable to be with her boy because she could not afford to miss work. Instantly, Mary's perspective shifted to one of compassion. She uses this teaching story to point out that we never know the complete story about other people—why they act as they do, or what befell them prior to an incident. Although this example represents a minor insult, it illustrates a tool that helps in even more extreme situations. Realizing that we cannot see the full picture and assuming that the individual is in some sort of need can quickly turn a simple clash into a blessed moment.

How often must we forgive? Jesus of Nazareth gives the answer in

Matthew 18:21-22. "Then Peter came to Him and said, 'Lord, how often shall my brother sin against me, and I forgive him? Up to seven times?' Jesus said to him, 'I do not say to you, up to seven times, but up to seventy times seven.'" Jewish law held that one was to forgive another up to three times. Probably expecting his master's commendation for his generous number of attempts, Peter learned there is no limit to the rightful number of times to forgive another.

Do you have someone to forgive? If you do, remember that you are not justifying the action the other person took. Nor are you putting yourself in harm's way again. The purpose of forgiveness is to remove a block in *you* that is keeping love from flowing through to you. Again—from *A Course in Miracles*—forgiveness is a shift in perception that removes a block in you to your awareness of love's presence. You are not doing this for the other person; you are doing it for yourself.

In the absence of feeling absolutely betrayed by someone, don't let yourself off the hook for exercising forgiveness. Although you don't immediately recognize grudges or even annoyances toward others, you might have some serious blocks that impede your flow of energy that you don't think of as resentment. This is often the case with my coaching clients, who seldom think of people they need to forgive. As we explore deeper, however, there are usually a handful of people who have disappointed them by their actions, e.g., being rude or by seeming to be neglectful. Whether or not our perception matches the other person's is irrelevant. The irritation sticks in our craw and saps our energy. At times, we blame others to escape accountability for our own circumstances. As the following story illustrates, we can even blame unlikely recipients in our lack of awareness.

Harold and Mabel

An elderly couple sat on their porch swing, waiting for guests to arrive to celebrate their sixtieth wedding anniversary. As they swayed back and forth, Harold said to Mabel, "You know, Mabel, I been thinkin'. Remember when we'd just been married a couple of years and the locusts came and destroyed all our crops?"

Mabel nodded.

"Well, Mabel, you were right there by my side!"

"And you remember, back in the '70s, when there was that awful flood that washed away our home?"

Mabel nodded.

"Well, Mabel, you were right there by my side!"

"And remember that terrible flu epidemic in the '90s when I got so sick I thought I was gonna die?"

Mabel nodded.

"Well, Mabel, you were right there by my side."

Mabel nodded.

"You know, Mabel, I been thinkin'. I believe you're bad luck!"

I enjoy this story because it humorously points to the human inclination of blaming unpleasant outcomes on someone or something outside ourselves. We then seem to relish carrying a grudge, protecting it, and nurturing it with additional justifications. With enough care, these resentments gain momentum, consuming energy we need to use productively for becoming our vision.

List five people toward whom you hold some resentment and begin your forgiveness work today. You will be amazed at the lightness you feel after actually letting resentment toward someone go—even if it takes time to accomplish the task.

_____

_____

_____

_____

Malleable vs. Fixed Personalities

Another widely accepted belief that must be uprooted is that we have "fixed" personalities. In the modern Western world, we have become obsessed with personality types. I have used personality tests extensively in leadership trainings. It can be helpful to identify your current dominant traits or determine where you need to shore up your personal skills to present yourself in a more positive light or to be open to other perspectives. However, damage occurs when you believe your traits are fixed—that they are part of your DNA and not subject to change.

Benjamin Hardy has written several outstanding articles in the blog "Medium" that offer extensive research data to show that personalities are formed, not birthed, and they can be reformed if desired. We come to the conclusion that we are a particular "personality type" based on our behaviors—not on our gene pool or on "permanent" characteristics. If I happen to like to work at night and sleep late in the morning, I probably would say I am a "night owl," not a "morning person." However, by changing my behavior for sixty-six days and arising early, following a set morning routine, I would incorporate that behavior into my automatic regime. Then I'd become uncomfortable *not* getting up early and not having my own personal quiet space to do whatever is important to me. Hence, I'd consider myself—after a relatively short span of time—a real "morning person."

Here is a paradox that works to our advantage. Throughout this book we have explored different ways our minds control our behavior. It is important to note also that the mind and body are inextricably linked. Therefore, the relationship works both ways. To change a mindset, it is often most expedient to incorporate a change in physical behavior at the same time, and the desired results will occur exponentially faster. We will see other examples in the next chapter of how performing a physical task while changing our thinking has a dramatic impact on our mental and emotional state. In regard to the personality, psychologist Benjamin Hardy says it best: "Personality isn't a fixed thing. Personality is DEVEL-

OPED. And it never has to stop developing." This a very important point, for, as author Darren Hardy says, "To achieve what you have not, you must become what you are not. You have to grow into your goals."

You must *be that person*. That is why we must always keep in mind that we are malleable, changeable, and transformable. William Durant says, "We are what we repeatedly do. Excellence then, is not an act, but a habit."

"Nothing in this world is constant but change," to quote one of my favorite ancient Greek philosophers, Heraclites—and that is the *good* news! The title of this book you are reading evolved from *Creating Your Own Dream Life* to *Becoming Your Dream* precisely to illustrate this point. I wanted you, my reader to understand from the beginning that change begins as an "inside job." By cutting away the mental, physical, and spiritual parts of us that do not align with our dreams and replacing them with empowering beliefs/models and habits/systems, we will *become* our dreams, magnetizing every aspect that is reflected in *who we are*. A significant part of this reformation involves discovering and uprooting false beliefs that sabotage our growth.

The JoHari Window, described by Joseph Luft and Harrington Ingram in 1955, gives us a good illustration. The window contains four quadrants, as follows:

1. The **first quadrant** is wide open—seen by ourselves and others—those aspects of ourselves that we recognize, and others see them clearly as well.

2. The **second quadrant** is that part of us that is known only to us. We mask these traits, keeping them hidden so that observers don't notice them.

3. In the **third quadrant** is our blind spot. Here we are not aware of characteristics that others notice in us. We might react negatively when they are pointed out to us, or we can calmly consider them and get feedback from other sources to determine

whether the observation is valid, and, if so, whether these are traits we want to change.

4. The **fourth quadrant** contains our hidden secrets, buried so deeply that even we are not aware of them. When they show themselves in unexpected moments, we are usually shocked, wondering where *that* came from. Hopefully, this quadrant is the tiniest one, for those buried emotions, memories, and stories can impact us tremendously. Carl G. Jung, noted Swedish psychologist, writes, "Until you make your unconscious conscious it will direct your life, and you will call it fate."

Do not settle for less than your ideal life because you have held on to destructive, limiting beliefs!

Epigenetics has burst forth with information in the last decade that now proves that even our genes don't have the final say, as they are on a "toggle switch" to be activated or not. Deepak Chopra and Rudolph Tanzi, PhD have produced in their recent work paradigm-breaking research, definitively showing that life choices made throughout our life—from early on—determine the activation of the particular genes we have at birth. In their book *Super Genes*, they "introduced the concept of DNA as something dynamic, ever changing, and totally responsive"[53] to the many body and mind choices we make through the years.

This model is in direct contradiction to the long-held belief that what you are born with is what you get. Chopra and Tanzi speak to the efficacy of the whole-system approach to address physical wellbeing. As I mentioned in Chapter 6, the mental and physical approach of our biofeedback program, by increasing personal awareness, automatically encompassed the whole emotional, physical, mental, and spiritual aspect of the patient. No longer were the isolated symptoms (e.g., headaches, back pain, irritable bowel syndrome) the focus of treatment as patients became conscious of themselves in new ways.

---

53 Chopra, Deepak and Rudolph Tanzi. *Super Genes*. p. 6.

This whole-system approach is not a new fad or buzz word; it is the total destruction of the false walls Western medicine has used in an attempt to isolate each body part, each cell. As Chopra and Tanzi point out in *The Healing Self,* nature does not recognize these human-made categories. They make no sense to the wide network of information highways circulating throughout the body at all times. In fact, these scientists and healers have educated us to recognize that the body's intelligence is far more advanced and millions of years older than that of our fairly young (in biological terms) thinking brain.

For that reason, we will occasionally refer to our system as the *bodymind* to help each of us remember that in order to make informed life choices, we must tune back in to the body's voice. Its many signals alert us to physical interruptions, and they go much further than that to let us know the truth or falsehoods being spoken and the intuitive responses we need to make in complex situations. These impulses guide, support, nurture, and allow us to live in life's present moment. Depending on our employment of the three factors listed below, there are powerful tools at our disposal:

1. Everyday experience
2. Simple lifestyle choices
3. Techniques to increase awareness

Adi Shankara, medieval sage and philosopher, declared that people get old and die because they see other people doing it. That sounds a bit far out, until one considers the rapid decline of elders who begin associating only with declining elders. In several studies cited in *The Healing Self,* simply introducing novel experiences reversed the aging process for the majority of people in the studies. This reinforces earlier statements about the importance of peak experiences. Recognizing that our choices and the subsequent habits that become part of our actions determine our traits is totally freeing!

As mentioned above, a *second type* of buried beliefs exists—those personal ideas about the "me" I think I am. Jung identified several versions of self: the outward self—that we display to the world, the conscious inner self by which we describe ourselves to ourselves, and the unconscious inner self—the part that resides so deeply in our subconscious that we're not aware of it. Yet this unconscious inner self has tremendous power in dictating our actions. By raising our awareness of our patterns and by critically listening to the inner dialogue running in our heads, we can replace these faulty beliefs with empowering ones.

Habits to Acquire

Since our beliefs become our actions and our actions become our habits, the next step is to closely examine the habits of our daily life. While models of the world can be changed relatively quickly by replacing faulty beliefs with updated ones, our systems or habits of how we do things require step-by-step processes to modify and then eliminate beliefs that no longer serve us. These systems are firmly ingrained and have worked well many times in the past—and that intermittent reinforcement makes them even stronger. A quick look at bamboo as a metaphor will illustrate this point because, perhaps not surprisingly, we have a lot in common with bamboo.

The Bamboo

The Chinese bamboo is a delicate plant that requires the proper soil acidity to prosper. When planted and nourished in the proper environment, a young bamboo sprout grows an inch the first year. The second, third, and fourth years, the sprout grows an additional inch each year. During the fifth year, the Chinese bamboo grows eighty feet! During the earlier growing periods, this clever plant was developing an extensive root system that could support its phenomenal upcoming height.

Your habits sometimes require steady nurturing to become automatic (fortunately not five years!). By performing the new supportive behaviors consistently, remember that you are building an extensive foundation that will easily handle your giant leaps when they occur. While many self-help texts tout twenty-one days of repetition as the typical length of time needed for a new habit to become automatic, research doesn't support that. The actual time varies, depending on the complexity of the habit being acquired, but the average is sixty-six days. Therefore, it is necessary to be patient as you build stronger neural networks in your brain. Nevertheless, sixty-six days is an extraordinarily brief period of time "in the overall scheme of things," considering the many years of an improved life that the newly automated habit will bring.

Routines to Adopt

Forming a morning routine allows you to establish firmly the new actions you want to take proactively that day. You are taking the first step for a quantum leap in productivity. Many people reach for their phone or email to check the incoming stream of information. Strategic planners, on the other hand, set aside their early morning time as "sacred space" to get themselves pointed in the direction they determine. They are proactive, not reactive, taking charge of their thoughts and bodily movements. As we reviewed in Chapter 6, there are countless variations to a personal morning routine, but the keys to a life-changing routine are as follows:

1. Allow yourself time for review of your "Why."
2. Prioritize next action steps. (Per Stephen Covey, schedule your priorities instead of prioritizing what's on your schedule.)
3. Elevate your mindset with inspirational reading, music, and writing.
4. Crystallize your intent through journaling.
5. Move all your body parts in healthy, healing ways in some form of exercise.

The elements that elevate your metabolism (a cold shower), bring you inner calm (a walk in Nature), or help you image your end goal in finite detail with deep feelings, are all beneficial, and can be added according to your situation. An old Zen proverb states that you should meditate for twenty minutes each morning, unless you have a particularly busy day ahead, and then you should meditate for an hour!

A morning routine's success also depends on the consistent practice of an evening routine of:

1. Reflection
2. Journaling
3. Previewing and prioritizing the next day's action steps
4. Planning for a good night's sleep

Habits to Eliminate

The habit of working longer hours and sleeping less stems from the competitive model discussed earlier. Science is now demonstrating what wisdom traditions have known for thousands of years: sleep is an essential restorative, healthy process. Body, mind, and spirit need the refuge of peaceful sleep to rejuvenate and recover from the many assaults of the day's activities, the environment's toxins, and the collected waste in our bodymind (to borrow the scientists' term).

Also needing recovery time is the subconscious, working 24/7, keeping the massive bytes of information organized and ready for recall when needed. Chronic diseases, such as cancer and diabetes (Type II), and digestive disorders gain their foothold when the immune system becomes overwhelmed by lack of restoration time. By working in short focused periods, like two or three hours at most, and then taking restful, relaxing breaks for an extended period, i.e., an hour, one can double productivity. We must rid ourselves of the notion that "being busy" is being productive.

The following story illustrates for us how we can chip away the beliefs we no longer need to get at the ones that do serve us.

The Statue

A woman cleaning a sculptor's studio became fascinated by a large block of marble in one of the empty rooms. Each evening when she arrived, she examined the large stone, and soon discovered that chunks were missing. Every day the marble lost some more pieces, until on one inspection, the cleaning lady determined it was a head. She eagerly rushed each night to the spot to check the progress. As she entered the studio early one evening, she encountered the sculptor, Gutson Borglum, who was putting some final touches on the bust. As the interested cleaning lady studied the head, she suddenly exclaimed, "Oh my! How did you know Mr. Lincoln was in there?" Borglum replied, "I just carved away everything that wasn't Lincoln."

This sculptor demonstrates for us precisely what we must do with our non-serviceable beliefs and habits. We must "carve away everything that isn't Lincoln," leaving only that which is our best self. Extraordinary people constantly review and update their systems, recognizing that how they do anything is how they do everything.

As the amazing entrepreneur Elon Musk says, "I think it's very important to have a feedback loop, where you're constantly thinking about what you've done and how you could do it better. I think that's the single best piece of advice: Constantly think about how you could be doing things better and questioning yourself." Deepak Chopra says, "Every time you are tempted to act in the same old way, ask if you want to be a prisoner of the past or a pioneer of the future."

As you recall, the meaning-making machine—our conscious brain—usually has done an unfiltered "brain dump" into the subconscious before we reach the age of seven. Our natural maturation process includes forming beliefs. As sponges, our young minds soaked up information at an astronomical rate as we crawled, walked, and then ran around the environment picking up details from all our sensory input. We had two tendencies as children:

1. To take things personally

2. To make assumptions

These mental habits made sense when we were children. Most things our parents said in our presence—whether intentional or not—*were* about us. Also—we needed to make some assumptions just to figure out what was going on in the big, complex human world we had recently entered. *But* (and this is the kicker) as we became older and entered a wider, more expansive environment, most of us kept right on assuming and taking things personally—even when the conversation or circumstance had *nothing* to do with us! Don Miguel Ruiz in *The Four Agreements* addresses both of these habits, demonstrating how destructive they are. He admonishes us, "*Don't take anything personally*. By taking things personally you set yourself up to suffer for nothing…. When we really see other people as they are without taking it personally, we can never be hurt by what they say or do. Even if others lie to you, it is okay. They are lying to you because they are afraid."[54] Ruiz goes on to explain that we all tell lies to others and to ourselves when we are trapped in our social masks. "When you make it a strong habit not to take anything personally, you avoid many upsets in your life. Your anger, jealousy, and envy will disappear, and even your sadness will simply disappear."[55] As Eleanor Roosevelt put it, "No one can make you feel inferior without your permission." Breaking this habit of personalizing feedback helps you to break many small habits and routines that trap you in emotional hell.

Regarding assumptions we make, Ruiz points out that we have so many things in our lives that need explanation, that are beyond our understanding or level of information, that our conscious brain fills in the gaps with possible answers—taken from the models of life stored in our subconscious. Not only are the assumptions possibly false, but also the beliefs on which the assumptions are based may be far from any reality. As we evolve, we must inspect these beliefs for truth and assess them for alignment with our core val-

---

54 Ruiz, Don. *The Four Agreements*. p. 57.
55 Ibid. p. 58-59.

ues. Isaac Asimov tells us, "Your assumptions are your windows on the world. Scrub them off every once in a while, or the light won't come in." When the master thinkers we have studied, such as Emerson, Thoreau, Gandhi, and His Holiness the Dalai Lama, spoke of "right thinking," they were explaining that these types of erroneous thinking must be overcome to raise one's consciousness. The rewards for this intensive work are great—happiness and freedom.

The Learning Journey

We all must go through the Learning Journey when we want to incorporate a new habit, or when we want to tackle a non-effective ingrained habit to eliminate. The Learning Journey consists of four stages:

1. We begin as an *unconscious incompetent*. We don't have the foggiest idea of what we don't know.

2. Then we become a *conscious incompetent*. We now know what it is we must learn, but we don't know how to do it.

3. Next is the *conscious competent* phase. We now know what we are learning to do, and we have acquired the skills to do it, but the task takes keen attention and focus to complete.

4. The final stage is the *unconscious competent*. We are so practiced and developed in the skill we are learning that it can be done "on automatic."

Learning to drive a car is an example often used to compare these stages in *acquiring* a habit. Initially, driving looks simple, but we have no idea about its actual complexity that we still must learn. Once behind the wheel, we become aware of all the skills we do not have to be able to drive successfully. After practice and persistence, we become more skilled in driving, but we must put our full attention on the road, the car, and our surroundings. Finally, we become so adept at driving that we sometimes surprise ourselves by arriving at our destination without much awareness of the journey.

All learners must pass through these stages, some taking longer at different points than others, but it is helpful to recognize that those things we might consider our weaknesses *can* be learned.

In the same manner, given the necessary motivation, urgency, and confidence, you and I can *replace* any ineffective habits with a stronger system. As the sculptor revealed, we must cut away any pieces that no longer serve us. This action is no different from replacing an outdated app on our phone that cannot accomplish our new goal, except it will take a bit longer. Be sure to give the new system sixty-six days to fall into the unconscious competent stage. The reward in productivity is immense, so excuses be gone!

In *Willpower Doesn't Work,* Benjamin Hardy adds a valuable step that is often overlooked in self-improvement formulas: the need to *change our environment* to match our vision. We do this in five steps:

1. We begin with clearly defining and testing our vision.

2. Then we reinforce our belief in our deserving the dream.

3. After gaining an understanding of the powerful, energetic being we are, and of the vibrational energy field we inhabit, we become aware of and learn to use our superpowers (HOTS) to our advantage.

4. We then learn to navigate the gap between where we are (circumstances) and who we are becoming.

5. To speed up the process dramatically, allowing for quantum leaps instead of incremental gains, Hardy inserts this action: We use our critical thinking skills (HOTS) to *align our environment with our goals.*

By actively changing our environment for successful achievement of our next steps, we incrementally increase our chances of success. As Ariana Huffington says, "Act as if life is 'rigged' in your favor." So get busy and rig your outer life in your favor. Since we are part of a continuous energetic field, it is critical that we eliminate distractors and add supports to the environments of all major areas of our lives. For example:

- Our workspace must support and stimulate creativity and productivity.
- Our environment must encourage overall wellness.
- Our relationships—both personal and work-related—must be aligned with our personal values.
- Our relationship with money—the way we earn it, spend it, save it, and give it must reflect our vision of a meaningful life.
- The way we prioritize our time in this world must concur with our vision of an ideal life.

By taking these steps, we are aligning our inner and outer games—bringing the A-game to both arenas.

You will achieve your dream life by following the steps we have covered so far: 1) raising your vibration to the level of your vision, and 2) culling the beliefs and habits that derail you.

By universal law you will become one with your vision, seeing results even beyond your wildest expectations! Before you can fully receive your bounty, you must, however, add another step:

3) You must learn to maintain this higher state of being as your dominant energy. Sustainability requires continued growth and continuity of awareness, a higher state of consciousness.

A Master Mind Group

- **HOTS ALERT: Intuition**

So how does one accomplish this ever-moving upward state? Do *not* attempt to do it alone! That you must "go it alone to succeed" is another fallacy of our individualistic society.

Hill found, as he researched the world's 500 wealthiest men at the time, that most relied on their community of excellent minds for their success. Figures such as Andrew Carnegie (who commissioned Hill to study these successful people) used Master Mind Groups to harness the collective power of the intelligent people around them. They found these benefits:

- Gaining access to Infinite Intelligence with the aid of creative imagination.
- Gleaning the accumulated experience of the entire group.
- Holding the dream in a safe "brain trust."

Henry Ford's success skyrocketed when he became close friends with Thomas Edison. A later alliance of these two minds with Firestone, Burroughs, and Burbank helped create an auto industry previously incomprehensible. When two minds together commit themselves to experiment and research new ideas, they create an intangible force, much like a third mind, that propels them forward. Social psychologist David McClelland of Harvard found from his research that your "reference group" can determine as much as 95 percent of your success or failure in life. Dr. Gail Matthews's research on high achievers showed that individuals with written goals were 39.5 percent more likely to succeed. The even more surprising results were that individuals who wrote their goals and sent them to an accountability partner were 76.7 percent more likely to achieve them!

As you proceed with your dream, find one or two trusted people with whom you can meet periodically to discuss progress and challenges. Be certain that your Master Mind has your best interest at heart, and that its members understand the universal principles you are using to become your dream. It's imperative that they believe in you, have a like-minded philosophy, and are willing to support you in stretching your talents.

You and I must each stay at the helm of our mind, assuming leadership of our most precious asset. We must also inspect our planned actions to be sure they align with our intentions since we recognize that our behavior yields our results.

What action steps can you take in the next month to align your environment with your vision? In the next week? Tomorrow?

_____

_____

_____

_____

What habit can you begin this week to improve your productivity toward your goal? How will you track your progress? Who will support you?

_____

_____

_____

_____

What habit can you eliminate this month that will make room for a better practice for your wellbeing? How will you rearrange your environment to accommodate and support this change? How will you track your progress? What positive results will you achieve? Who will help you with accountability?

_____

_____

_____

_____

What will become your most effective morning routine? By when will you have it implemented? Have you calendared it?

_____

_____

_____

_____

What are you going to set up as your most efficient, useful evening routine that will complement carrying over unfinished work/goals? By when will you implement this routine? Is it on your calendar?

_____

_____

_____

_____

"I dream my painting, and then I paint my dream."

— Vincent Van Gogh

PART IV

RECEIVING YOUR DREAM

The following chapters will put in easy-to-follow, concrete terms, additional positive steps that will help to prevent your succumbing to the gravitational pull back to the old familiar paradigms that have kept you stuck. They will include your individual practice of gratitude, of happiness, of expanding your abundance, and of giving and receiving.

Throughout my years of coaching, regardless of the age, gender, or purpose of the client, one fact has remained: For a transformation to be permanent, the person making the small, transitional changes must step wholeheartedly into the being he or she is working to become. The person must seek friendships and associations with like-minded people. He or she must structure the environment to facilitate a new way of being, such as stocking healthy foods, arranging the work environment to enhance focus, etc. If these steps are not taken, the gravitational pull back to the old, familiar habits becomes daunting. Josh Waitzkin describes in great detail this mental and physical shift in his book *The Art of Learning*. He explains that the new habits must be practiced with such focus and regularity that they fall into the subconscious mind. Here "Lizzie" takes over and runs the new behaviors at incredible speed, without using the energy of the conscious-thinking mind, which is now freed to find fresh solutions to new situations.

Andrew Newberg, in hundreds of fMRIs on patients and research subjects, has demonstrated that actual structural changes occur in the brain as one makes new routines habituated. Even small, transitional changes form new neural pathways that can be developed into frequently used circuits.

These points are relevant to us as we continue our upward spiral journey. Our progress begins with small steps. Then the momentum accelerates, and events occur with synchronicity. As this new ideal life materializes, it becomes increasingly important to move our dominant mental vibratory pattern upward. Love and above must become our frequent experience. The upcoming chapters will show us how to move to new altitudes.

"What you leave behind is not what is engraved in stone monuments, but what is woven into the lives of others."

— Pericles

CHAPTER 9

LIVING YOUR GRATITUDE AND HAPPINESS

HOTS ALERT: Will, Perception, Memory

"Gratitude is not only the greatest of virtues,
but the parent of all others."

— Marcus Tullius Cicero

The house reminded me of a Girl Scout camp cabin in its simplicity. The night was bitterly cold, and I had shut off the water before I left town. Standing in total darkness with chilling winds battering me, I had to figure out a way to break the ice off the heavy water meter lid.

As I stood there shivering, feeling more isolated than ever before, my spirit descended so low that I thought, *So this is what it's like. This is it. I have totally hit bottom.*

The engulfing despair was deeply painful; I wasn't sure I could endure it.

I finally cranked open the heavy, icy iron lid, turned on the water, dragged my luggage inside, and fell into bed without even brushing my teeth. Sleep would not come despite my fatigue. I lay, heartbroken, on my bed and stared at the plain walls until the sun rose.

I had worked hard to achieve my dreams. After putting myself through college, I'd earned graduate degrees and achieved my vocational goals. For what? These accomplishments had vaporized, as had some of my finances. I loved my family with all my being, yet my husband was gone, my home was gone, my business was gone, and my extended family had disappeared. It seemed there was nothing the locusts had not destroyed in the seven years following my husband's death.

Traveling on the red-eye from San Diego, I had just returned from visiting my two daughters. During a pizza feast with my girls and many of their friends the previous night, the conversation had taken a funny turn. Funny to everyone else. My girls and I often found ourselves laughing jovially together at our missteps in life, cleverly labeling them to de-energize them. That night—New Year's Eve— the witty quips from my daughters had brought tears of laughter from the group—to all, that is, except me. My perception was that my daughters considered me a dismal failure as a mother.

In the early morning dawn, despite my despondent mood, a quiet thought popped into my head. *Be thankful for your abundance.* I knew deep down I would only find relief through gratitude. True thankfulness seemed impossible; I was filled with a scarcity mindset. I wanted to scream out in pain.

Pushing aside my pain and self-doubt, I found the strength to ask myself, "What do I love about my life?" At first, I could not stir up any authentic grateful emotion. Finally, I landed on the thought of my pillow. "Yes! I love my goose-down pillow." It snuggled me and caressed me and held me safely as I slept. "I love my pillow." As I repeated these thoughts/words, I felt a huge shift. Something was different. A wave of combined relief and—dare I say it—yes, a hint of joy crept into my being. Immersed in these feelings, I noticed a slight movement at my bedroom window.

There, perched in the new morning sunlight, was the most radiant red cardinal I had ever seen. In this stark, cold, icy weather, he was a total surprise. I don't know how long he stayed, but I do know he put on a glorious performance, just for me. Fanning his crest to display his prowess, Mr. Cardinal turned and looked directly at me. He proceeded to stretch his wings and turn slightly to each side, as if Geppetto were jiggling his strings. A few dewdrops from the melting ice on his crest and wings glistened as if stage lights shone on him.

Soon, in spite of myself, I chuckled at his prancing state. He was fully aware of his own beauty and of the joy he brought to the world.

After my front row VIP seat experience of the cardinal, it became easy to think of other blessings:

- My eyes that could behold him
- My arms and face that could feel my wonderful pillow
- The breath that flowed in and out, bringing me life
- The amazing winter landscape that surrounded me as those magnificent trees built up their energy to grow thousands of new leaves in the spring
- The dried leaves that made rustling sounds as a squirrel scampered by
- My ears that could detect the delightful sounds

The list of blessings kept growing, and I began writing them down lest my awareness dimmed. The boldest gift came when I reflected upon the loving faces of my two daughters. Our lives had been fraught with some difficult circumstances, but many of those trials we had weathered together by laughing and steadying each other. As I reviewed and revisited the "Joan roast," I became deeply thankful we could laugh and play together in an atmosphere of unconditional love.

When I shifted my perspective from myself as a victim to that of a co-creator, I recognized my preposterous assumptions. Instead of their mocking me, my daughters had been celebrating the common bond of humanness we share. The moment I made room for gratitude in my heart, love flowed in and healed my flawed perception—raising my vibration to become one with the glorious life that surrounded me.

Have you ever experienced a similar despair? I hope not, but since life must have its valleys for us to be able to see the mountains, it is likely you have—under different circumstances. The best solution to this depth of feeling works most effectively if used early in the process. Remember these two questions to interrupt the patterns evoked by that secret, hidden quadrant mentioned earlier that (no matter how tiny it is) can pull you off your upward movement and send you spiraling downward:

1. Do I know this story to be 100 percent true?

2. What do I know that challenges the story and broadens my perspective?

I share this not-so-proud time of mine with you because it glaringly illustrates that the questions above—if applied at *any* time during my melt-down, but ideally *immediately* when my feelings of "worthless mother" stuck their noses up for air—would have saved the day. How different the flight home could have been, as I might have reflected on the many times, particularly when recounting goof-ups, we three had laughed together! I had all the necessary information to disqualify any of my horrors, but instead, I gave them center stage, as if masochism were my nature. We can *choose* how we want to react to situations. Unfortunately, we don't always use this miracle of choice.

Many stories of the human journey find a climax in the perspective shift from victim to creator. New empowerment creates a mental and emotional opening for new ideas to flow in, and a call for action removes the horrible "stuckness" of despair. Psychologists Robert A. Emmons and Michael McCullough elaborate on the empowering effect of being genuinely thankful: "Gratitude promotes optimal functioning at multiple levels of analysis—biological, experiential, personal, relational, familial, institutional, and even cultural. I continue to notice the societal shift now since most acclaimed books written about organizational leadership devote a great deal of space to cultivating a sense of gratitude among corporate members at all levels.

Stephen Karpman, MD, synopsized the roles we play during almost every internal and external conflict (e.g., marital, friendship, home, work, around the world) as the Karpman Drama Triangle. In his model are the three roles of victim, rescuer, and persecutor. Depending on the circumstances, each of us can play any of the roles, and as the conflict escalates and continues, a player can shift among the roles.

Victim: The world has harmed me. Life isn't fair. I am powerless, oppressed, and victimized.

Persecutor: "It's all your/their/the economy's fault. I wouldn't have done it if you hadn't started it." This is the critical, blaming voice sitting in absolute judgment.

Rescuer: This is our enabler piece, swooping in to "fix" the situation. This role wants to rescue the other person, avoiding letting him or her take responsibility, or it wants to stop the internal voices, distancing oneself from any conflict or uncomfortable feelings.

Lyssa Danehy deHart in *StoryJacking: Change Your Inner Dialogue/ Transform Your Life*—a "must read" for anyone wanting better clarity on inner thinking—illustrates how the stories we tell ourselves control our lives. In Chapter 34: Along Came TED, deHart points out the significant shifts in perspective that can alter one's role in any conflict. She extols David Emerald's work in this area, saying, "The beauty of TED is that it simply makes sense of the Dreaded Drama Triangle (DDT) by way of a fable."[56]

In *The Power of TED: The Empowerment Dynamic*, Emerald offers an effective anecdote to the DDT as follows:

56 deHart, Lyssa Danehy. *StoryJacking.* p. 251.

The Victim shifts to **Creator**:

Your shift in this role is in recognizing that you can choose your responses—that you are greater than circumstances when you focus on your vision and purpose. As the Creator, you ask, "How would the person of my dream act in this situation?"

The Persecutor shifts to **Challenger**:

Instead of being the angry interrogator, you are now curious about how the other person is going to act, allowing her to make her own choice. You let go of any attachment to the choice of the Creator.

The Rescuer shifts to the **Coach**:

As a Coach does, in this role you lose the Rescuer/Enabler mindset, shifting to a perspective of empowerment for both you and the Creator. You see each of you as whole, creative, capable, and resourceful. You support the other person on her journey.

So what do these dynamics have to do with our feelings of gratitude and happiness? Everything—since our interactions with other people have such dramatic impact on our emotions and vice versa. When we learn to shift our perspectives in such dramatic ways, the inevitable plot twists in life don't derail our gratitude for our relationships or our happiness in living.

In *The Code of the Extraordinary Mind*, Lakhiani points to a gratitude practice as the leading happiness boost. He lists seven other scientifically proven benefits of gratitude:

1. More energy
2. More forgiving attitudes
3. Less depression
4. Less anxiousness
5. More feelings of being socially connected
6. Better sleep
7. Fewer headaches

Practice seeking more gratitude every day. Dr. Paige Tabor, author, chiropractor, and health coach, shares her perspective:

> True health has many components such as what we eat, how frequently or intensely we exercise, our total movement and mobility each day, how effectively we manage stress, and the number of hours we sleep. How we feel day to day can be impacted by the chemicals we are exposed to in personal care products, laundry detergent, and food containers. The quality of our food—either laden with pesticides and raised with added hormones or completely clean—can make a difference in whether or not we are robust. There are many practical action steps we can take daily to improve our health.

> That said, I firmly believe the first step to true health, the *most critical step*, is developing a grateful attitude. Having a spirit of gratitude and working *with* the resources we have access to instead of feeling antagonistic toward the challenges we face is a powerful way to promote our own vitality and make positive leaps toward a place of true health.

> Think of three things you can be grateful for right now. Say aloud those three things. Expound on why you are grateful for them. Notice how you feel as you focus on blessings in your life. Being grateful is a practice, a choice, and a powerful step toward true and lasting health.

Let's pause now to complete the exercise Tabor recommends. List three things you are grateful for right now.

_____

_____

_____

_____

Flood your entire body, your breath, your very consciousness with appreciation. Your dominant vibration will elevate, slowly at first, and then incrementally. As your gratitude practice continues and grows, it will assume a self-fulfilling aspect, for you will naturally notice much more to make your heart sing with thanks.

> "I find that the more willing I am to be grateful for the small things in life, the bigger stuff just seems to show up from un-expected sources, and I am constantly looking forward to each day with all the surprises that keep coming my way."
>
> — Louise Hay

Happiness

> "To be happy, make others happy."
>
> — The Dalai Lama

As Rabbi Spinoza taught centuries ago, I have found gratitude to be the first cousin to happiness, for a grateful heart opens itself to receive joy. In a study by Emmons and McCullough, subjects who daily wrote down five things they were thankful for showed a 25 percent difference in happiness from a second group who wrote down five negative things from the week.[57]

Most people feel that success brings happiness, but actually, the opposite is true. From a sustained joyful state, you will encounter opportunities and "lucky breaks and positive coincidences" due to an expanded awareness. Shawn Achor, author of *The Happiness Advantage* (I'm a huge fan of this scientist of positive psychology), says, "It turns out that our brains are literally hardwired to perform at their best not when they are negative or even neutral, but when

---

57 Emmons, Robert A. and Michael E. McCullough. "Counting Blessings Ver-sus Burdens: An Experimental Investigation of Gratitude and Subjective Well-Being in Daily Life." p. 377-389.

they are positive. Yet in today's world, we ironically sacrifice happiness for success only to lower our brain's success rates."[58]

Jeff Olson, in his book *The Slight Edge*, gives a list of five happiness habits that Shawn Achor teaches to the members of Olson's organization:[59]

1. Each morning, write down three things you're grateful for. He teaches to select three different things each day in different areas of your life.

2. Journal for two minutes a day about one positive experience you've had over the past 24 hours.

3. Meditate daily. Watch your breath go in and out for about two minutes to train your mind to focus.

4. Do a random act of kindness each day. A specific act at the same time each day, i.e. sending an email to someone thanking or acknowledging them, can help consistency.

5. Exercise for 15 minutes daily. Simple cardio or a brisk walk is an effective mood elevator.

Olson includes tips from other happiness researchers:[60]

- Make more time for friends (this has been proven the most effective longevity predictor).
- Practice savoring the moment.
- Practice having a positive perspective.
- Put more energy into cultivating your relationships.
- Practice forgiveness [as we covered in Chapter 8].
- Engage in meaningful activities.
- Practice simple acts of giving [as we'll cover in Chapter 10]

Olson adds his own recommendation:

---

58 Achor, Shawn. *The Happiness Advantage*. p. 129.
59 Olson, Jeff. *The Slight Edge*. p. 104.
60 Ibid. p. 105.

- Read at least ten pages of a good book daily. [I heartily concur, though I prefer at least fifty!]

Einstein said the most important decision one must make is whether or not this is a friendly universe. With the assumption that we live in a friendly, giving, abundant world that is just as eager to give you what you want as you are to receive it, you will have all of your HOT Skills on the alert, scanning for the bounty destined for you!

Your *imagination* will detect new opportunities for you that pessimists might miss, or it will even create entirely new ideas for you.

The still, small voice of *intuition* will drop crumbs for you to follow as you make strides toward your dream.

The *will*, unhampered by negative distractions, will remain focused on the necessary target of your vision.

Memories of pleasant past events, restructured, rescripted memories, as well as future memories now crystallized in the mind, will gain momentum in a friendly, cooperative milieu.

Perception, when shifted to the bright side of life, will paint a bright picture of progress, possibility, and probability.

Uncommon Reason will assure you that you are an inseparable part of all creation, having a bit of God in you with unlimited potential.

You will notice that once you make the commitment to be happy about your many blessings, your HOTS will allow you to discern that your current conditions are simply what your ideal world looks like as it is coming together. A deep sense of happiness lets you enjoy the beautiful little sparkles along the way as you take action to become your brilliant dream.

The following story illustrates that if we keep an open mind, the challenges we face can prove to be of great benefit to us. Likewise, the victories we score might have pitfalls. Therefore, instead of determining an experience to be good or bad, it is wise to keep aligning our progress with our dream.

The Farmer and His Son

One day a farmer and his son were on their way home from market, when their ox fell over dead in the road.

"Oh, no! This is terrible!" cried the son. "Our ox is dead. How will we ever till the fields?"

"Who's to know what's good, and who's to know what's bad," replied the farmer.

The next fall, after carefully tending the crops by hand, the farmer and his son had a bumper crop.

"Oh, this is grand," shouted the son. "We will have money to spare!"

"Who's to know what's good, and who's to know what's bad," replied the farmer.

They took their crop to market, and with their proceeds bought a fine horse. On the way home, the boy, who was riding, fell off the horse, and broke his leg.

At home with a leg splinted and suspended in traction, the son wailed, "Oh no! This is terrible! I have a broken leg and am in pain. What a horrible time!"

"Who's to know what's good and what's bad," replied the farmer.

A week later, the army came through recruiting every able-bodied man to go to war. Of course, the son was not eligible. "Oh—this is great! I have been spared from going to war and can mend safely at home!" exclaimed the boy.

The story continues, but you get the idea. Instead of acting like Tarzan gripping his strong vine and swinging wildly from anguish to ecstasy, we must moderate our reactions—using the Stoics' wisdom. By remaining calm, seeking moderation, and walking centered in our path, we will avoid the extremes of reactionary moves and allow some of the otherwise overlooked blessings of our current circumstances to emerge. Josh Waitzkin, chess and martial arts master, states this idea clearly: "I believe an appreciation for simplicity, the everyday—the ability to dive deeply into the banal and discover life's hidden richness, is where success, let alone happiness, emerges."[61]

Often we see people chasing happiness, as if she were some fleeting wood nymph. They race after new toys, exotic food and drink, thrilling games, and many other pleasures. Don't get me wrong here—these are all fun, but they don't bring *sustained* happiness. The truth is that happiness is an inside job. A study that brought this fact home to me was done by Dan Gilbert, psychologist and author of *Stumbling on Happiness*. Gilbert collected data on people who won the lottery and on paraplegics. His astounding findings were that one year later, both groups were equally happy with life! There is an "impact bias" that says that situations occurring more than three months ago have less intensity and duration than more recent events. How this amplifies Adam Smith's declaration, "Tis nothing good or bad but thinking makes it so."

Brendon Burchard, renowned business and performance coach, issues bracelets to his clients that read, "Bring the Joy." His directive is to be the person of increase who brings happiness to the scene. Some of the happiness triggers he suggests are:

> *Doorways*—When entering a room, say the mantra silently, "I bring joy to this room and am open to serve."

> *Notification Trigger*—Set an alarm label in your phone to notify yourself with a happy reminder. Burchard uses "Bring the JOY!"

> *Waiting Trigger*—Whenever in line waiting to purchase something or make an inquiry, check in with yourself to determine your energy level. Are you LOVE and ABOVE? Do you feel the energy you want to radiate out into the world? Instead of tapping your foot impatiently, enjoy life and feel vibrant while you wait!

---

61 Waitzkin, Josh. *The Art of Learning.* p. 187.

Touch Trigger—When you greet someone, gently touch her or him, or hug him or her. Human touch is a necessary element for our wellbeing, physically, mentally, emotionally, and spiritually.

Gift Trigger—Make it a practice to give gifts to others. These gifts can be small or large, a compliment, a flower, a word of appreciation, a note of congratulations, or a loving glance. When you receive a gift, be sure to receive and acknowledge it fully, saying with enthusiasm, "What a gift!"

Stress Trigger—Choose an internal calm amid the chaos around you. Remember that stress is self-created, so choose not to engage. Ask what positive thing you can do to help the situation.

In addition to the happiness triggers, Burchard advocates journaling as an important practice. By recording each evening the things that made you happy during the day, you will put in your memory a positive tapestry of events. Without writing down and cementing the happiness inside you, there will be a tendency to forget these moments and recall only those that brought negative emotion that needed attention. I don't make a practice of reading my journals, but occasionally I'll glance back—particularly at those end-of-day thoughts that brought me joy. The review of these happy moments brings me as much joy—if not more—as I received initially. When I discovered this pleasant memory jogger, I began journaling my nighttime reflections in a separate little book that's special to write in, for they hold such power. At times, I even rewrite a memory to happen the way I would have liked it to go. Research shows we overwrite our memories based on newly acquired information all the time anyway—so why not take advantage of that hack and become happier about an interaction with another person?

Vishen Lakhiani advocates three systems or daily habits that bring happiness:

1. **Practice gratitude** (see above)

    An additional shift in perspective related to practicing gratitude that you will want to employ he calls "Appreciating the Reverse

Gap." When you become discouraged by noticing the gap between your current position and the dream person you are becoming, *reverse* your attention to the point from which you started. Instead of dwelling on how far you think you have to go, look at how far you have already come since the beginning of your journey. This reverse gap—from past to present—can offer great encouragement versus the present to future gap.

2.   **Practice forgiveness** (see Chapter 8)

Every person who has wronged us—even slightly—must be forgiven to liberate ourselves to unblock our flow of happiness.

3.   **Practice giving.**

Giving is one side of a two-sided single coin. It is the side that elevates your happiness and creates ripples that impact many you'll never know. As a school principal, I was often shocked at notes from students or parents thanking me for the smallest gesture I had made for a student—often years earlier. A simple sincere compliment, encouragement, or caring action can bring a person's mental state to a whole new positive level. Lakhiani says it best, "Be merciless with your kindness." You can give time, love, understanding, compassion, ideas, skills, wisdom, energy, physical assistance, admiration, praise, recognition, attention, and so many other aspects of yourself. Albert Schweitzer echoed this advice: "I don't know what your destiny will be, but one thing I know: The only ones among you who will be truly happy are those who will have sought and found how to serve."

Receiving is the other side of the same coin. Giving and receiving are both part of the Law of Circulation. One cannot be sustained without the other. If you give a lot, but receive only a little, or if you receive a lot but give only a little, the balance of nature is lost. Many think receiving is such an easy task, but it is actually one of the highest acts of appreciation. To reject or impede the gifts that come to you creates a block in the flow of energy.

We each have access to the ocean of abundance that the universe offers, yet often we come to the sea with only a thimble to fill. The lesson is to rejoice in both giving and receiving, allowing happiness to encompass both the positive and negative poles of energy that create perfect balance in our lives.

Another of Lakhiani's coined terms that I appreciate and use is *Blissipline*, for it beautifully implies that the art of being happy is a regular daily discipline available to each of us. Lakhiani says to live in Blissipline. "Extraordinary minds understand that happiness comes from within. They begin with happiness in the now and use it as a fuel to drive all their other visions and intentions for themselves and the world." After all, when we ask the key question, "What would I *love*?" are we not actually asking, "What would make me truly *happy*?"

There are other very effective happiness hacks available. These hacks do not offer sustained happiness, but they provide many happy moments for the fully present and aware person. These blissful moments then translate into a string of happy times, which transform into a positive mental state.

Again—remember our mantra: *Love and above!*

By feeding our sensory input and HOTS with many joyful jogs of the mind, heart, and soul, we develop an overall sense of wellbeing and joy in being alive. Try some of the joyful jogs on the list below and add your own items with your own flavors of happiness, giggling, and fun. Then share them with us all at *BecomingYourDream.com*.

- Visiting a doggy park to stand in the sea of unconditional love between dogs and their owners
- Watching young children play
- Walking in nature
- Spending time outdoors looking closely at the many colors, shapes, textures, and sizes of living things
- Playing music that speaks to your soul
- Dancing as if no one is watching

- Singing with all your heart
- Telling a dear one you love him or her
- Feeding all your senses with their delights
- Breathing deeply! Breathing deeply again! Rejoicing that you are alive!

We have examined and employed the steps to bridge the gap between our current circumstances and our vision. Now it is time to reap the harvest of abundance that awaits you.

"Many wealthy people are little more than janitors of their possessions."

— Frank Lloyd Wright

CHAPTER 10

RECEIVING YOUR ABUNDANCE

HOTS ALERT: Reason, Intuition

"The key to abundance is meeting limited circumstances with unlimited thoughts."

— Marianne Williamson

The universe we live in is an abundant environment. Nature provides richly for each of us. We perceive scarcity, but this view is a flawed one—a problem of competition instead of cooperation—not one of limited resources. As business coach T. Harv Eker says, "Don't believe everything you think."

A good example of how we can allow our thoughts to limit us comes from the elephants.

River Reed Thinking

In India, elephants are used instead of cattle as beasts of burden. While a calf is still young, one of its legs is chained to a stake in the ground. No matter how much the baby struggles, he cannot break free. Finally, he stops trying. Eventually, a rope made of woven river reeds replaces the heavy chain, but the elephant, now large enough to easily break loose, makes no attempt to escape.

Similar to these elephants, each of us has had massive programming from our culture, parents, siblings, educators, religious figures, and peers. "Money doesn't…," "You can't have everything…," "They (those rich people) don't care about…," "Money is…," "Money is not…" You can fill in the blanks because you've heard these or similar phrases delivered to you with emotion by respected people expressing their beliefs (which, by the way, were also implanted in them with "river reed" logic).

The flawed mass concept of money that grips so many of us comes from a "me or them" society. Either I get all I can now, or they'll get too much, and there will be none left for me later. This idea becomes an automated belief that controls our actions and our results with our money transactions.

Early in my financial planning career, I read a doctoral thesis that stated a person's belief and "money story" can actually impact the outcome of that individual's money accumulation. I thought the hypothesis strange, but the grad student presented interesting evidence, so I tucked the information away with a question mark over it.

In the ensuing ten years of my practice, I was surprised, at times even shocked, to observe that thesis playing out. Although I gave the same concerted attention to each client, I still found that the naysayers made self-sabotaging decisions regarding their accounts, or some other financial aspect of their lives, despite my cautionary advice. These observations began my quest to learn more about the spiritual aspect of money. It wasn't, however, until I began my layman's study of quantum physics that the phenomenon made sense.

At the time, I believed people's actions were "good" or "bad" in a moral sense, and that money correlated positively with the person using it. The evidence, however, did not support my beliefs. As an example, one of my dearest friends, a doctor, philanthropist, and genius, held a solid belief that he would lose money on every endeavor he tried. Despite brilliant work in several fields, including a thriving medical practice, he always somehow managed to offset his financial gains. All areas of his life flourished—except in the area of monetary wealth. He was loved and admired by all of us who knew him, but his bank account did not reflect the abundant being he actually was. His keen interest and enthusiasm in many subjects prompted him to develop businesses outside his medical practices that drew money like a magnet, yet I watched in bewilderment as he repeatedly created wealth and subsequently destroyed it. His genius and loving spirit were evident, but his firm belief that he always lost money became a self-fulfilling prophecy.

There are universal truths about the currency of money. Money can seem mysterious when it shows up "like magic." Let me tell you a personal story to illustrate this point.

Money Magic

I was thrilled to be going away to college, to learn, to build new relationships, and to escape the discord in my home environment. Confident that my education was secured by my grandmother's generosity, I discovered—just prior to my departure—that two family betrayals had resulted in no college funds. Fortunately, late in the summer, I secured a $200 tuition scholarship, and a friend agreed to drive me to West Texas to attend Texas Tech.

Arriving in Lubbock, I quickly obtained three jobs (in a men's clothing store on the college strip, as a dorm switchboard operator, Lily Tomlin style, and as a grader/tutor for the math department). Living frugally, I met my expenses until the spring semester because my work hours had decreased over the holidays. Our dorm mailboxes had glass fronts, so the daily pink slip I received, showing I was delinquent in paying my room and board, was clearly visible. Each day, I hurried back from morning classes to snag it, but not before others noticed I was delinquent. As I privately fretted about how to earn extra money to pay my deficit (credit cards, as we now know them, were non-existent), I was shocked one day to open my mailbox and find in place of the dreaded pink slip, a paid receipt for my room and board. (In the early '60s, dorm fees were $65 per month.)

I stood there baffled as a long line of girls shuffled past me to the cafeteria. Suddenly, a girl who lived down the hall from me (whom I barely knew) stood before me, looking a bit flushed. She quickly explained that to pay my dorm fees, thirteen girls had united to donate blood at the local blood bank. They were each paid $5.00, and they made an impromptu party out of the juice and cookies they were given afterward.

I was dumbfounded, for I had shared my problem with no one. My new friend assured me that each of them received more joy from their gift than I could know. She said college was highly competitive on an academic and social scale, but their celebration of kindness and giving had given them a new perspective.

I had to expand my own grace to accept their gift. This was one of my lessons in understanding that giving and receiving are two sides of the same coin. The ebb and flow of life is interrupted if the giver cannot express her generosity for another being, or if the receiver cannot accept the gift with thanks.

In my sophomore year, my textbook bill exceeded my available funds. Three weeks into the semester, I was really concerned about being so far behind without textbooks to study by. "Out of the blue," a representative from Dunlap Department Store called to say I'd been awarded the freshman beanie fund donation. "For what?" I asked. He explained that their store donated the proceeds from the sale of freshman beanies, the official school cap traditionally worn by all freshmen their first semester, to a deserving scholar. How my name was entered, let alone selected, was a mystery, yet when the donation equaled exactly the cost of my textbooks, I knew the gift was from my Source. I learned through these and other experiences what money could provide, and to feel the love that flows with money when given with a generous heart. I also developed an inner knowing that this intelligent universe always provides.

Since my college years, I have discovered that money itself is neutral, but it becomes charged with the energy around its movement. As "currency," it flows best through the most open passageways, for circuits with the least diversity have the most strength. Money resembles water in that its influence can penetrate the tiniest or the grandest of spaces. It becomes involved in almost every relationship. Like water, money can be a force for great good or great evil.

If these ideas initially seem foreign to you, test out some of the following practices, and watch your flow of abundance in the form of money increase.

1. Appreciate the money you have now (reread Chapter 9 if needed).

2. Call on your *will* to adopt the firm attitude that money is plentiful.

3. Treat money well when it comes to you, regardless of the amount. Don't stuff it blindly into your pocket or purse, but welcome it into your possession with joy and with a comfortable, honored, holding spot until you use it, or allow it to become a money magnet, drawing more in kind.

4. Celebrate the monetary gains of those around you and in your global community. Bless their money. The evidence of their increase indicates that you, too, are a person capable of increase.

Great humanist and futurist Buckminster Fuller, in his 1970s series of lectures entitled "Integrity Days," observed that for centuries we lived with a scarcity mentality, thinking we had to compete and fight for everything we needed to survive. He said this perception may have been true much earlier, but by the 1970s, we had reached a point in which we could make and do things so much more easily that it was, by then, entirely possible for everyone everywhere to sustain a reasonably healthy and productive life. He declared this was a turning point for humankind where we could move from a "you *or* me" to a "you *and* me" world, where each of us can meet her or his fundamental needs to live a fulfilling life.

Fuller predicted that this transformation would take fifty years and would require our financial systems to change drastically. He spoke of the changing world view that emerged after the first manned lunar landing by the Apollo II crew in 1969. Seeing planet earth from the moon, mankind had its first view of our planet as a whole. Its beauty and fragility stunned us. In her seminal book *The Soul of Money*, Lynne Twist postulates that this vision was the beginning of our global community.

Twist speaks of world hunger as a mystery, for currently there is enough food on the earth to feed everyone several times over. However, one-fifth of humanity is hungry and malnourished. Farmers are paid not to raise food. Cattle consume enough food to feed every hungry child and adult. Waste abounds.

Lynne Twist's mission to end world hunger has uncovered some astonishing facts:

The problem that causes world hunger is *not*, as we might believe:

- Lack of caring
- Poor distribution
- Politics
- Lack of money

The problem is a scarcity mentality. For example, the many massive aid efforts after crises in Ethiopia, Somalia, Biafra, and Cambodia provided short-term relief, but not long-term solutions as intended. The *fundamental assumption of scarcity* undermined every effort despite the economic circumstances.

By empowering people to "author their own recovery," and by uncovering the myth and lie of scarcity, Twist illustrates that the Hunger Project has opened avenues of possibility and hope.

While I've found great insights in Twist's authentic experience with world hunger issues, it is her "journey into money and soul" that makes her book an excellent read for anyone wanting to better understand his or her relationship with money. The author articulates beautifully my deepest beliefs, observations, and inspirations about money. She, too, has been moved by the struggle many of us have had with money.

When I want to help coaching clients become aware of the obstacles preventing transformations in their lives, I ask them to observe their money flow. This flow reveals their focus of attention, even in areas outside the financial realm.

What is my point? Aligning your money with your soul values means inspecting and changing (if needed) each of the following:

- Your life's priorities
- What you think is real
- Your lifestyle
- How you earn money
- How you invest money

In a world where there is enough, we can all thrive, not at each other's expense, but through cooperation and collaboration.

Lynne Twist shares the views of many of us that this is the time "to hospice with compassion" the death of old, outworn, unworkable systems. New constructs can be midwifed for sustainable systems and new ways of being.

The same mental faculties that we use to accumulate things of depletion to "puff ourselves up" can be used to express our deep love for people and our affirmations of life.

Here we see *intuition* at work as each of us listens to the still, small voice inside that shares our soul's longings. We also find *reason* actively calculating solutions as we take action to fulfill our passion.

However, let me be clear. This shift in perspective is *not* a call for paucity or scarcity in the giver's own life. We are meant to be richly abundant, following Nature's lessons. Furthermore, we can only give at the level of our own self-worth.

It is critically important to use money to obtain high quality, to be discriminating about the beauty with which you surround yourself. This discretion, of course, rules out purchases to impress others; only purchases to satisfy your inner values can actually bring you sustained pleasure.

The irony between the old paradigms we must release and the new beliefs about money to embrace is that being mindful—selective and purposeful—in buying for yourself and for your family and friends *increases* the flow of money. It is not necessary—and, in fact, unadvisable—to downgrade or censor your longings. Instead, up-leveling opens up your perspective to new possibilities!

The difference is in one's basic mindset: one of gratitude and joy in receiving plus a firm conviction of Nature's bounty for all, rather than one of greed and consumption combined with a fear of scarcity. While we see today many of the latter scenarios, a global shift is occurring—an awareness that we are all connected and we all share in the personal responsibility to care for our beautiful living planet.

Twist gives a parallel in Nature of these contrasts, taken from evolutionary biologist Elisabet Sahtouris. A caterpillar, at certain life stages, becomes a voracious, over-consumptive glutton, consuming hundreds of times its own weight. The more it consumes, the more sluggish it becomes. At this point in evolution, the imaginal cells (minority specialized cells) connect with each other to become the genetic drivers of the caterpillar's metamorphosis. The over-consumptive caterpillar becomes the nutritive soup for the emerging butterfly.

Using this metaphor, Twist describes our present time, when "imaginal cells" are taking responsibility for the transformation of our world. The visionary thinker otologist Werner Erhard notes that "Transformation does not negate what has gone before; rather it fulfills it. Creating the context of a world that works for everyone is not just another step forward in human history; it is the context out of which our history will begin to make sense."

Let's look at a story that exemplifies the compounding effect that money has when given in a loving, compassionate vibration.

A Time of Need[62]

Many years ago two boys were working their way through Stanford University. Their funds got desperately low, and the idea came to them to engage the famous pianist Ignacy Paderewski for a piano recital. They would use the funds to help pay their board and tuition.

The great pianist's manager asked for a guarantee of $2,000. The guarantee was a lot of money in those days, but the boys agreed and proceeded to promote the concert. They worked hard, only to find that they had grossed only $1,600.

After the concert the two boys told the great artist the bad news. They gave him the entire $1,600, along with a promissory note for $400, explaining that they would earn the amount at the earliest possible moment and send the money to him. It looked like the end of their college careers.

"No, boys," replied Paderewski, "that won't do." Then tearing the note in two, he returned the money to them as well. "Now," he told them, "take out of this $1,600 all of your expenses, and keep for each of you ten percent of the balance for your work. I will be satisfied with the amount remaining."

The years rolled by—World War I came and went. Paderewski, now premier of Poland, was trying to feed thousands of starving people in his native land. The only person in the world who could help him was Herbert Hoover, who was in charge of the U.S. Food and Relief Bureau. Hoover responded and soon thousands of tons of food were sent to Poland.

After the starving people were fed, Paderewski journeyed to Paris to thank Mr. Hoover for the relief he had sent to him.

"That's all right, Mr. Paderewski," was Hoover's reply. "Besides, you don't remember it, but you helped me once when I was a student at college, and I was in trouble."

62 H. Jackson Brown, Jr. and Rochelle Pennington. *Highlighted in Yellow*. p. 70-71.

This story illustrates Dr. Martin Luther King's sentiment that life's most urgent question is: What are you doing for others?

Lynne Twist challenges us with these sentiments, matching those of my heart, asking that our love, our heart, our word, and our humanity be reflected in our earning and use of money:

> [U]se the money that flows through our life…to express the truth and context of sufficiency.

> …to hold money as a common trust that we're all responsible for using in ways that nurture and empower us, and all life, our planet, all future generations.[63]

This doctrine of sufficiency, generosity, and prosperity is the perfect segue to using money to expand our purpose. As we initially move from conditions of scarcity and a compromised lifestyle, it is important and necessary to focus on our individual longings and discontent to learn what our soul wants to express. As the early changes in our dominant mindset bring new results that comfort and satisfy us, our vision grows to include our community, both local and global.

Some of us have the passion to touch all humanity, to end world hunger, to provide clean water for all, to see that education is accessible for each child. Others are drawn to conservation of our beautiful planet's natural resources, while still others want to focus on the flourishing of birds, animals, fish, and insects.

What is important to remember is that *there is no singular good*. The individual genius of each of us is needed and must be brought to the task. You might ask, "How will the world benefit if my dream is to find my soul mate?" By finding a person to share your love and to enrich your life, you will become a source of joy for all those around you. As a couple, the two of you can demonstrate the power of a deep loving connection.

The difference each of us makes in this world is illustrated in the following story.

---

63 Twist, Lynne. *The Soul of Money*. p. 257.

The Starfish Story

A young man is walking along the ocean and sees a beach on which thousands and thousands of starfish have washed ashore. Further along he sees an old man, walking slowly and stooping often, picking up one starfish after another and tossing each one gently into the ocean.

"Why are you throwing starfish into the ocean?" he asks.

"Because the sun is up and the tide is going out, and if I don't throw them further in, they will die."

"But, old man, don't you realize there are miles and miles of beach and starfish all along it? You can't possibly save them all. You can't even save one-tenth of them. In fact, even if you work all day, your efforts won't make any difference at all."

The old man listened calmly and then bent down to pick up another starfish and threw it into the sea.

"It made a difference to that one."

How beautifully this story illustrates someone's understanding that there is no singular good. We are all connected in the field of life. Each starfish can lead to a healthier ocean (they are the most important predator in the shallow water ecosystem), just as our effort to light one candle can lead to a much brighter world.

Remember our mantra: *Love and above*. By living at a higher dominant vibration, you will raise the vibration of those around you.

History teaches us that great world-changers can come in small packages.

Mother Teresa raised millions of dollars and nursed tens of thousands of people.

Martin Luther King, Jr. led a peaceful revolution for civil rights of all humans.

Dr. Victor Frankl demonstrated that a human can endure the worst atrocities—even those of a concentration camp—and still prevail to teach and inspire millions.

As a six-year-old boy, Abraham Lincoln took his dying mother's words to heart, "Be kind to each other." He overcame many failures and prevailed to end the practice of slavery in the United States.

Nelson Mandela was imprisoned for twenty-seven years; eighteen of those were spent in horrible conditions of starvation and grueling work, on the isolated rock quarry on Robben Island. He continued to fight for a democratic and free society while in prison. At age seventy-one, Mandela helped negotiate the end of apartheid, and then served one term as president of South Africa.

These words, taken from Mandela's speech after his release from years in prison, hold true today:

"The need to unite the people of our country is as important a task now as it always has been. No individual leader is able to take on this enormous task on his own." (Feb. 11, 1990)

Still focusing on Mandela's message to unite, today we must substitute the phrase "our country" for "our world," because our problems have become so complex that they require global solutions. No singular country has the ability or capacity to come up with or to implement all the answers. A collaborative effort, not a competitive one, is required to bring about the changes that will save and enrich our planet.

Not one of us, but all of us together, living in our abundance, giving of our abundance, and sharing our talents can make this planet become our dream world. As your own rich life grows, filled with good health and wellbeing, loving relationships, and time and money freedom, your meaningful purpose will also expand—as Life ever-seeking to express itself more fully. As my coach David Simon put it so beautifully: "A wave of individuality rises from the unbounded ocean, and for a time, forgets that it is the ocean in disguise. When the wave begins looking inside, the memory of wholeness is rekindled, and the wave again knows itself as unbounded, infinite, and eternal."[64] As each of us

64 Chopra, Deepak and David Simon. (2012). *21-Days of Inspiration*. p. 38.

grows in strength as a wave, let us remember to look inside and come together as the one ocean that we are. Harbhajan Singh Yogi teaches, "If you can't see God in All, you can't see God at all." As you learn to receive your goodness and your abundance with grace and gratitude, you will find that the same inner divine light that shines so brightly in those you love is also present in each of us. You will be joyfully motivated to share your wealth of knowledge, resources, experience, and happiness with the world as a whole. There is no private good. Your gifts are part of the fabric of humanity.

Now our collective consciousness *must* rise into a spirit of cooperation. The only way this new awareness can come about is from a groundswell of people living consciously, using their special talents to make our planet whole.

If you feel my sense of urgency, I'm glad. We are living in an energetic, vibrational world that, I believe, wants us to rise above the lower vibrations of guilt, fear, anger, and violence. The tools are here. A critical mass is forming as more and more people from all walks of life are raising their consciousness to the collective good.

My daily mantra is "Love and above"—which implies raising my mental state to the vibrational level of love (calibrated at 500—as explained in Chapter 4). I haven't yet developed the skills to sustain this elevated mental state, but I make it my practice to choose to visit it frequently, and I invite you to do the same. The "love level" is a joyful place, and it is so strong that it nullifies surrounding lower energies. Isn't that a promise for each of us to pursue for the benefit of our successors?

Knowing that your best song is still in you, continue your work to become what life wants to express through you. In the process, your candle will become a torch that will light thousands of others. As you reach a plateau, the ever-upward spiral of Life's growth will beckon you again, just as we learn from the ancient script of the Talmud that each blade of grass has an angel over it whispering, "Grow! Grow!" Tiny sprigs of grass will push up through cement to get to the light. Let us—each one of us—employ our special gifts to do the same.

"And the day came when the risk to remain tight in a bud was more painful than the risk it took to bloom."

— Anais Nin

A FINAL NOTE: EXPANDING YOUR GAME

The first takeaway of this book was that you have subjective superpowers, Higher Order Thinking Skills (HOTS), that will, when consciously directed, affect your objective world. Next, you designed your dream. After testing it from several angles, you were encouraged to break the inhibitory chains of your preconditioning and to upgrade your vision. You were nudged to go ahead and dream full-out.

After reviewing a bit of quantum physics to be able to understand how transformation works, you looked at the power of your thoughts, words, and feelings as they impact your world.

You found in the next chapter the importance of movement, and you learned highly strategic moves to anchor each day to be productive and to move you closer to your vision. You learned thought patterns to overcome the inevitable challenges you would encounter on your hero's journey.

Next, you unpeeled some of the layers of mental fog to search out beliefs you had acquired that no longer fit your new wider perspective. You also examined the habits that needed pruning for your growth to be able to move from changing in incremental stages to leaping in quantum bounds after removing the blocks restraining it.

You learned that, in order to be sustained, success must be enveloped with gratitude, even before it comes. You found that thankfulness is actually a prerequisite for your putting on and becoming your new reality. Then, as a grateful receiver, you found your awareness expanding to the supreme joy of giving. Initially, you found happiness in giving to family and friends. Eventually, as you became one with all that is, your dream grew even larger, encompassing the world.

It has long been my passion to demystify the mystical by giving scientific explanations for phenomena that many of us humans could not understand until the technology was developed to measure physically what was going on subatomically. After all, we've been trained to trust only our five senses for our interpretation of reality, and if we could not see it, we just weren't going to believe it, no matter how magical the promise seemed. Some people, such as my grandmother, were able to absorb and believe the spiritual laws, to actually apply them with faith that the laws were precise, and to look to the happy results for their own confirmation. Many of us, however, have had such predominant skepticism that we want more evidence before attempting "miracles." My sincere wish is that these pages have inspired you to use these highest gifts from your Creator. Hopefully, you now understand that the Universe wants you to be supremely fulfilled even more than you want it. By continuing to study and by applying ancient wisdom in the context of what science is discovering about reality, you and I can understand and use the principles that govern life.

I implore you to stop simply wishing and *become your dream*. As you arrive at that plateau in your ever-upward moving spiral, you will then look out upon a new vista, a vista grander than you could have imagined even a year ago. If you have received the message of this book, you now understand that you are limitless. As you continue to co-create with your divine Source, you will enrich Life for all of us.

"Your time is limited, so don't waste it living someone else's life."

— Steve Jobs

REFERENCES

Throughout *Become Your Dream*, I have quoted directly from some books and articles. Other books and articles listed below have influenced my own philosophy and coaching practice. Each one listed has helped form the principles found in *Become Your Dream*. I share them all with you, whether they've shown up directly or indirectly in my writing.

Achor, Shawn. (2010). *The Happiness Advantage: The Seven Principles of Positive Psychology That Fuel Success and Performance at Work.* New York: Crown Publishing.

Allen, James. (n.d.) *As a Man Thinketh.* Camarillo, CA: DeVorss Publications.

Amen, D. G. (2000). *Change Your Brain, Change Your Life: The Breakthrough Program for Conquering Anxiety, Depression, Obsessiveness, Anger, and Impulsiveness.* New York: Times Books.

Barker, Raymond Charles. (2011). *The Power of Decision: A Step-by-Step Program to Overcome Indecision and Live Without Failure.* New York: Penguin Group.

Behrend, Genevieve. (1951). *Your Invisible Power: The Mental Science of Thomas Troward.* Camarillo, CA: DeVorss and Company.

Brown, H. Jackson, Jr., and Rochelle Pennington. (2001). *Highlighted in Yellow: A Short Course in Living Wisely and Choosing Well.* Nashville, TN: Rutledge Hill Press.

Burchard, Brendon. (2017). *High Performance Habits: How Extraordinary People Become That Way.* Carlsbad, CA: Hay House.

_____. (2014). *The Motivation Manifesto: 9 Declarations to Claim Your Personal Power.* Carlsbad, CA: Hay House.

Brown, Brené. TED Talk. "The Power of Vulnerability." June 2010.

Chew, Luis. Medium blog.

https://medium.com/personal-growth/warren-buffets-5-25-rule-will-help-you-focus-on-the-things-that-matter-2c383e09d13c

Chine, Kara. Team Better blog. https://blog.teambetter.com/something-scary.

Chopra, Deepak and David Simon. (2012). *21-Days of Inspiration*. Carlsbad, CA: The Chopra Center.

Chopra, Deepak and Tanzi, Rudolph E. (2018). *The Healing Self: A Revolutionary New Plan to Supercharge Your Immunity and Stay Well for Life*. New York: Harmony Books.

deHart, Lyssa Danehy. (2017) *StoryJacking: Change Your Inner Dialogue, Transform Your Life*. New York: Aviva Publishing.

Edison, Thomas Alva and Henry Ford. (2004). *The Edison & Ford Quote Book*. Fort Myers, FL: Edison and Ford Winter Estates.

Emmons, Robert A. and Michael E. McCullough. "Counting Blessings Versus Burdens: An Experimental Investigation of Gratitude and Subjective Well-Being in Daily Life." *Journal of Personality and Social Psychology*. Vol. 84.2 (2003): 377-389.

Emoto, Masaru. (2006). *Water Crystal Healing: Music and Images to Restore Your Well-Being*. Hillsboro, OR: Beyond Words Publishing.

Goddard, Neville. (2012). *The Power of Awareness:* includes *Awakened Imagination*. New York: Penguin Books.

Hardy, Benjamin. Medium blog. https://medium.com/@benjaminhardy.

_____. (2018). *Willpower Doesn't Work: Discover the Hidden Keys to Success*. New York: Hachette Books.

Hardy, Darren. (2010). *The Compound Effect: Jumpstart Your Income, Your Life, Your Success*. Philadelphia, PA: First Vanguard Press.

Hawkins, David R. (2002). *Power vs. Force: The Hidden Determinants of Human Behavior*. Carlsbad, CA: Hay House.

_____. (2012). *Letting Go: The Pathway of Surrender.* Carlsbad, CA: Hay House.

Hawkins, David R. and Pauling, Linus. (1973). *Orthomolecular Psychiatry: Treatment of Schizophrenia.* New York: W. H. Freeman & Co.

Hill, Napoleon. (1997*). Keys to Success: The 17 Principles of Personal Achievement.* New York: First Plum Printing.

_____. (2005). *Think and Grow Rich.* New York: Penguin Books.

Holliwell, Raymond. (1964). *Working with the Law: 11 Truth Principles for Successful Living.* Camarillo, CA: DeVorss and Company.

Keller, Gary with Jay Papason. (2012). *The One Thing: The Surprisingly Simple Truth Behind Extraordinary Results.* Austin, TX: Bard Press.

Kelly, Matthew. (2016). *Resisting Happiness: A true story about why we sabotage ourselves, feel overwhelmed, set aside our dreams, and lack the courage to simply be ourselves...and how to start choosing happiness again!* Erlinger, KY: Beacon Publishing.

Koch, Richard. (2019). *The 80/20 Principle: The Secret to Achieving More with Less.* New York: Doubleday.

Lakhiani, Vishen. (2016). *The Code of the Extraordinary Mind: 10 Unconventional Laws to Redefine Your Life & Succeed on Your Own Terms.* New York: Rodale.

Lipton, Bruce H. (2008). *The Biology of Belief: Unleashing the Power of Consciousness, Matter & Miracles.* Carlsbad, CA: Hay House.

Maltz, Maxwell. (1970). *The Magic Power of Self-Imaging Psychology: The new way to a bright, full life.* New York: Pocket Books.

Mickelson, Christian. (2017). *Abundance Unleashed: Open Yourself to More Money, Love, Health, and Happiness Now.* Carlsbad, CA: Hay House.

Murphy, Joseph. (2010). *The Power of Your Subconscious Mind:* Includes the Rare Bonus Book *How to Attract Money.* First Jeremy P.

Tarcher/Penguin Edition. New York: Penguin Books.

Newburg, Andrew, MD and Mark Robert Waldman. (2016). *How Enlightenment Changes Your Brain: The New Science of Transformation*. New York: Penguin.

Olson, Jeff. (2013). *The Slight Edge: Turning Simple Discipline into Massive Success & Happiness*. Lake Dallas, TX: Success.

Pagan, Eben. (2018). *Opportunity*. n.p.: Go Meta Publishing.

Pert, Candice B. (2003). *Molecules of Emotion: The Science Behind Mind-Body Medicine*. New York: Scriber Publishing.

Pritchett, Price. *You²*. (n.d.) Dallas, TX: Pritchett Printing.

Ruiz, Don Miguel. (1997). *The Four Agreements: A Toltec Wisdom Book*. San-Rafael, CA: Amber-Allen Publishing.

Schpancer, N. "Overcoming Fear." *Psychology Today*. Sept 20, 2010.

Sincero, Jen. (2017). *You Are a Badass at Making Money: Master the Mindset of Wealth*. New York: Penguin Random House.

Spiro, Mary. (2015). *7 Game-Changing Traits for Uncommon Success*. New York: Penguin Random House.

Twist, Lynne. (2017). *The Soul of Money: Transforming Your Relationship with Money and Life*. New York: W.W. Norton Company.

Waitzkin, Josh. (2007). *The Art of Learning: An Inner Journey to Optimal Performance*. New York: Free Press, 2008.

Wirt, Sherwood Eliot and Kersten Beckstrom. (1982). *Topical Encyclopedia of Living Quotations*. Minneapolis, MN: Bethany House Publishers.

Wattles, Wallace D. (2007). *The Science of Getting Rich: The Proven Mental Program to a Life of Wealth*. New York: Penguin Books.

Wolf, Fred Alan. (1989). *Taking the Quantum Leap*. New York: Harper & Row.

"Life should not be a journey to the grave with the intention of arriving safely in a pretty and well preserved body, but rather to skid in broadside in a cloud of smoke, thoroughly used up, totally worn out, and loudly proclaiming, 'Wow! What a ride!'"

— Hunter S. Thompson

ABOUT THE AUTHOR

Joan McManus is a business and leadership coach, published author, educator, and keynote public speaker.

As a teacher, Joan worked with students in college, secondary elementary, and preschool. Having a strong desire to investigate how the human brain learns, she earned a Master's degree in speech and language acquisition and became a board-certified speech/language pathologist. Working in both medical and educational settings, she treated adults and children.

A college background in mathematics led Joan to pursue an intensive three-year training program to become a certified financial planner. As an entrepreneur, for more than a decade Joan coached individuals and families in their financial decisions.

Returning to the field of education, Joan served as a teacher trainer. Following additional graduate study, she became a school principal, using her extensive experience in education, special needs populations, and finance. Following retirement from the school district, Joan continued her study of psychology and brain research, and became an avid student of quantum physics. Recognizing the immense physical, emotional, mental, and spiritual benefits of meditation, she also became a certified Primordial Sound Meditation instructor through the Chopra College.

Continuing her passion throughout her professional career for coaching others to live a happy and fulfilling life, Joan studied under Mary Morrissey and the Life Mastery Institute team to become a certified life coach and a life mastery consultant. Joan has inspired and supported many clients and students through her coaching, writing, and speaking. Joan enjoys traveling and spending time in nature with her husband, two daughters, extended family, and friends. She delights in releasing the hidden artist in her as she paints. Her mission is to help others empower themselves to become the people they dream of being, to live lives they love, being, doing, having, and giving everything they want.

ACKNOWLEDGMENTS

I deeply appreciate the coaches I've had over the last decade for giving me many new insights and strategies for growth. Thank you to Mary Morrissey, CEO of Life Mastery Institute, with whom I've studied extensively, and to Dr. Kirsten Wells, Dr. Deepak Chopra, Dr. David Simon, Brendon Burchard, Benjamin Hardy, and Eben Pagan. Each of these coaches has helped me with the ongoing process of becoming a better version of myself. I am so grateful to have been in the company of these masters.

So many outstanding authors have shaped my approach to life and to my coaching. Some whom I have relied upon heavily for this book are Dr. Wayne Dyer, Darren Hardy, Dr. David Hawkins, Napoleon Hill, Raymond Holliwell, Vishen Lakhiani (who, as CEO of MindValley, has brought so many great thinkers to my attention), Dr. Bruce Lipton, Don Miguel Ruiz, Lynne Twist, and Wallace Wattles. Dr. Deepak Chopra and Mary Morrissey are daily inspirations to me with their prolific writings. Thank you to these great writers and to the many others who have traveled with me through life in their books.

I so appreciate the members of my own family, especially my daughters, Kara Chine and Paige Tabor, who continually inspire me through their loving support and their own amazing personal growth.

Many loyal friends—Carol Joy Ackles, Betsy Birkett, Constance Faith, Skye Fitzkee, Candy Glade, Dana Hawkins, Gloria Hester, Holley Hood, David Kaatz, Lynn Kitchen, Julie Maitland, JoAnn McKinzey, Anne Prinz, Ouida Rowland, and Summer Smith—have supported me in this book-writing process.

My publishing coach, Patrick Snow, has been with me through the entire journey—from getting the first word of that "book in my head" committed to paper to the eventual book launch—helping me get my message out to you. His positive encouragement and sustained support have been invaluable.

My editor, Tyler Tichelaar, CEO of Superior Book Productions, has turned the rough spots into coherent passages, while keeping the integrity of my style intact. Tyler's patience, attention to detail, and sense of humor are characteristics that made the otherwise tedious process of rewriting a labor of love I really enjoyed.

Nicole Gabriel has made this book a pleasure to look at, as she applied her artistic eye and intuitive sense of meaning to my cover and gracefully arranged the text layout so that you, the reader, would enjoy reading it. Nicole continues to amaze me with her talent for creating beauty, even in her use of space.

My publisher, Susan Friedmann, CEO of Aviva Publishing Co., has offered valuable tools to market my book.

Additionally, my marketing coaches, including Jeff Walker, CEO of Product Launch formula, Ryan Levesque, CEO of The Ask Method, and Robert Martinez, CEO of Manifestation Technologies, have helped me immensely to get this book from my desk into your hands.

Saving my biggest supporter, cheerleader, and fan until last, I celebrate my husband, C.J. Without his encouragement and wizardly organizational skills, I might still be sitting on the floor, surrounded by hundreds of books, a myriad of articles, and reams of unnumbered pages, with a dazed look on my face. I give him my humblest thanks for the daily—yes, daily—nudge to move forward with my passion, and for his unlimited faith in me to produce a book that will speak to others to help them find their own greatness.

And thank you, my reader, for investing your precious time and energy into reading *Becoming Your Dream*. I hope you will absorb the primary point:

You have right now the mental skills to learn to synchronize your dominant mental attitude with the higher frequencies of the ideal person you dream of becoming. By using that gift of choice that is already yours, you will, by universal law, become your dream.

BOOK JOAN MCMANUS TO SPEAK AT YOUR NEXT EVENT

Joan McManus is an entertaining motivational keynote speaker, having worked with large and small groups to inspire growth and leadership. She provides highly interesting, thought-provoking talks, designed to engage and prompt the audience to action. Her talks can be scheduled for twenty minutes, thirty minutes, or an hour.

Joan also provides highly interactive workshops for people interested in business development, personal growth, and productivity. Combined with humor and compassion, she guides participants to form crystal-clear visions of their dream businesses. After testing their goals to escape the confines of conditioned thinking, each member of the group then forms his or her next action steps toward transformation. Every business owner leaves the workshop with a clear plan of action that includes focus, team-building, leadership skill enhancement, and sustaining momentum. The participants will also develop strong individual morning routines to anchor each day in productive action toward their visions.

To learn more about how Joan can help your group or organization, contact her at:

Joan@JoanMcManus.com
JoanMcManus.com
214-793-3203

"You have power over your mind—not outside events. Realize this, and you will find strength."

— Marcus Aurelius

URGENT!

BECOMING YOUR DREAM

Thank you for Reading My Book!

I love getting your feedback!

Your comments, your insights, your reactions are music to my ears.

I need your help to make the next edition of this book and my upcoming

related coaching course better.

Please leave me an honest review on Amazon letting me know what you thought of the book.

A HUGE SLICE OF GRATITUDE COMES FROM ME TO YOU FOR YOUR INPUT.

THANK YOU!

JOAN C McMANUS

www.ingramcontent.com/pod-product-compliance
Lightning Source LLC
Chambersburg PA
CBHW060013210326
41520CB00009B/870